W9-CUZ-009

CONCEPTS
OF PHYSICS

Alessio Mangoni

ISBN: 9798675717668

Books belonging to this series:
"Complex numbers", 9798674312185;
"Special relativity", 9798675703647;
"The mathematics of quantum mechanics", 9798645275037;
"The Dirac equation", 9798666724644;
"Relativity, decays and electromagnetic fields", 9798663840200.

DR. ALESSIO MANGONI, PHD
Scientist and theoretical particle physicist, researcher on high energy physics and nuclear physics, author of many scientific articles published on international research journals, available at the link:
http://inspirehep.net/author/profile/A.Mangoni.1

https://www.alessiomangoni.it

I edition, August 2020

Contents

II Special relativity

III The mathematics of quantum mechanics

IV The Dirac equation

V Relativity, decays and electromagnetic fields

Introduction

This collection brings together the five books of the series "concepts of physics". The books cover the following topics:

- complex numbers;

- special relativity;

- the mathematics of quantum mechanics;

- the Dirac equation;

- relativity, decays and electromagnetic fields.

These are basic concepts of physics, indispensable for its complete understanding.

Part I

Complex numbers

Introduction

This book is dedicated to complex numbers. In the first part we expose the theory, where the main arguments are: complex numbers in cartesian, polar and exponential forms, the complex conjugate, the modulus, properties and theorems about complex numbers, polynomial equations. In the second part we propose and solve some original exercise on the topics covered in the first part, to fix the concepts.

Chapter 1
General form

The general form of a complex number, $z \in \mathbb{C}$, is

$$z = x + iy, \qquad (1.0.1)$$

where x and y are real numbers, $x, y \in \mathbb{R}$ and they are called the real part and the imaginary part of the complex number z, respectively. The imaginary unit i has the following property

$$i^2 = -1.$$

The real and imaginary parts are denoted with

$$\mathrm{Re}(z) = x$$

and

$$\mathrm{Im}(z) = y.$$

The form of Eq. (1.0.1) is called the cartesian form of z, but there are also other forms to express a complex number, as we will see later.

If the imaginary part is null, $y = 0$, the number $z = x$ is real, while if the real part is null, $x = 0$, the number $z = iy$ is called a purely imaginary number.

Chapter 2

Elementary operations

2.1 Addition

Consider the two generic complex numbers in cartesian form

$$z_1 = x_1 + iy_1$$

and

$$z_2 = x_2 + iy_2 \,,$$

we can write

$$z_1 + z_2 = x_1 + iy_1 + x_2 + iy_2 = x_1 + x_2 + i(y_1 + y_2) \,,$$

so that

$$\begin{aligned} \mathrm{Re}(z_1 + z_2) &= x_1 + x_2 \\ &= \mathrm{Re}(z_1) + \mathrm{Re}(z_2) \end{aligned}$$

and

$$\begin{aligned} \mathrm{Im}(z_1 + z_2) &= y_1 + y_2 \\ &= \mathrm{Im}(z_1) + \mathrm{Im}(z_2). \end{aligned}$$

2.2 Subtraction

Consider the two generic complex numbers in cartesian form

$$z_1 = x_1 + iy_1$$

and

$$z_2 = x_2 + iy_2,$$

we can write

$$z_1 - z_2 = x_1 + iy_1 - (x_2 + iy_2) = x_1 - x_2 + i(y_1 - y_2),$$

so that

$$\begin{aligned} \mathrm{Re}(z_1 - z_2) &= x_1 - x_2 \\ &= \mathrm{Re}(z_1) - \mathrm{Re}(z_2) \end{aligned}$$

and

$$\begin{aligned} \mathrm{Im}(z_1 - z_2) &= y_1 - y_2 \\ &= \mathrm{Im}(z_1) - \mathrm{Im}(z_2). \end{aligned}$$

2.3 Multiplication

Consider the two generic complex numbers in cartesian form

$$z_1 = x_1 + iy_1$$

and

$$z_2 = x_2 + iy_2\,,$$

we can write

$$
\begin{aligned}
z_1 z_2 &= (x_1 + iy_1)(x_2 + iy_2) \\
&= x_1 x_2 + ix_1 y_2 + ix_2 y_1 + i^2 y_1 y_2 \\
&= x_1 x_2 + ix_1 y_2 + ix_2 y_1 - y_1 y_2 \\
&= x_1 x_2 - y_1 y_2 + i(x_1 y_2 + x_2 y_1)\,,
\end{aligned}
$$

so that

$$
\begin{aligned}
\mathrm{Re}(z_1 z_2) &= x_1 x_2 - y_1 y_2 \\
&= \mathrm{Re}(z_1)\mathrm{Re}(z_2) - \mathrm{Im}(z_1)\mathrm{Im}(z_1)
\end{aligned}
$$

$$(2.3.1)$$

and

$$
\begin{aligned}
\mathrm{Im}(z_1 z_2) &= x_1 y_2 + x_2 y_1 \\
&= \mathrm{Re}(z_1)\mathrm{Im}(z_2) + \mathrm{Re}(z_2)\mathrm{Im}(z_1)\,.
\end{aligned}
$$

$$(2.3.2)$$

2.4 Division

Consider the two generic complex numbers in cartesian form

$$z_1 = x_1 + iy_1$$

and

$$z_2 = x_2 + iy_2 \, ,$$

with $z_2 \neq 0$, we can write

$$
\begin{aligned}
\frac{z_1}{z_2} &= \frac{x_1 + iy_1}{x_2 + iy_2} = \left(\frac{x_1 + iy_1}{x_2 + iy_2} \right) \left(\frac{x_2 - iy_2}{x_2 - iy_2} \right) \\
&= \frac{(x_1 + iy_1)(x_2 - iy_2)}{(x_2 + iy_2)(x_2 - iy_2)} \\
&= \frac{x_1 x_2 + iy_1 x_2 - ix_1 y_2 - i^2 y_1 y_2}{x_2^2 - i^2 y_2^2} \, ,
\end{aligned}
$$

from which

$$
\begin{aligned}
\frac{z_1}{z_2} &= \frac{x_1 x_2 + iy_1 x_2 - ix_1 y_2 + y_1 y_2}{x_2^2 + y_2^2} \\
&= \frac{x_1 x_2 + y_1 y_2 + i(y_1 x_2 - x_1 y_2)}{x_2^2 + y_2^2} \\
&= \frac{x_1 x_2 + y_1 y_2}{x_2^2 + y_2^2} + i \left(\frac{x_2 y_1 - x_1 y_2}{x_2^2 + y_2^2} \right)
\end{aligned}
$$

so that

$$\mathrm{Re}\left(\frac{z_1}{z_2}\right) = \frac{x_1 x_2 + y_1 y_2}{x_2^2 + y_2^2}$$
$$= \frac{\mathrm{Re}(z_1)\mathrm{Re}(z_2) + \mathrm{Im}(z_1)\mathrm{Im}(z_2)}{\mathrm{Re}(z_2)^2 + \mathrm{Im}(z_2)^2}$$

and

$$\mathrm{Im}\left(\frac{z_1}{z_2}\right) = \frac{x_2 y_1 - x_1 y_2}{x_2^2 + y_2^2}$$
$$= \frac{\mathrm{Re}(z_2)\mathrm{Im}(z_1) - \mathrm{Re}(z_1)\mathrm{Im}(z_2)}{\mathrm{Re}(z_2)^2 + \mathrm{Im}(z_2)^2}.$$

Chapter 3
Complex conjugate

The complex conjugate of a complex number

$$z = x + iy\,,$$

is denoted with z^* or $\overline{z}$ and has the form

$$z^* = x - iy\,. \qquad (3.0.1)$$

Note that the complex number z^* has the same real part but opposite imaginary part of z,

$$\mathrm{Re}(z^*) = \mathrm{Re}(z)$$

and

$$\mathrm{Im}(z^*) = -\mathrm{Im}(z)\,.$$

3.1 Properties

The modulus of the sum of two complex numbers $z_1, z_2 \in \mathbb{C}$ coincides with the sum of their moduli, in fact

$$
\begin{aligned}
(z_1 + z_2)^* &= (x_1 + iy_1 + x_2 + iy_2)^* \\
&= \left(x_1 + x_2 + i(y_1 + y_2)\right)^* \\
&= x_1 + x_2 - i(y_1 + y_2) \\
&= x_1 - iy_1 + x_2 - iy_2 \\
&= z_1^* + z_2^*.
\end{aligned}
$$

Similarly, the modulus of the product of two complex numbers $z_1, z_2 \in \mathbb{C}$ is equal to the product of their moduli, in fact

$$
\begin{aligned}
(z_1 z_2)^* &= \left((x_1 + iy_1)(x_2 + iy_2)\right)^* \\
&= \left(x_1 x_2 - y_1 y_2 + i(x_1 y_2 + x_2 y_1)\right)^* \\
&= x_1 x_2 - y_1 y_2 - i(x_1 y_2 + x_2 y_1) \\
&= x_1 x_2 - (-y_1)(-y_2) \\
&+ i\left(x_1(-y_2) + x_2(-y_1)\right) \\
&= (x_1 - iy_1)(x_2 - iy_2) \\
&= z_1^* z_2^*.
\end{aligned}
$$

3.2 Useful relations

By adding the Eq. (1.0.1) to the Eq. (3.0.1) we obtain

$$z + z^* = x + iy + x - iy = 2x = 2\operatorname{Re}(z)$$

from which

$$\operatorname{Re}(z) = \frac{z + z^*}{2}.$$

Similarly, by subtracting Eq. (3.0.1) from Eq. (1.0.1) we have

$$z - z^* = x + iy - (x - iy) = 2iy = 2i\operatorname{Im}(z)$$

and hence

$$\operatorname{Im}(z) = \frac{z - z^*}{2i}.$$

Chapter 4
Modulus

The modulus of the complex number

$$z = x + iy \,,$$

is defined as

$$|z| = \sqrt{x^2 + y^2} = \sqrt{\mathrm{Re}(z)^2 + \mathrm{Im}(z)^2} \,. \qquad (4.0.1)$$

The squared modulus of z can be written also as

$$|z|^2 = zz^* = (x + iy)(x - iy) = x^2 - i^2 y^2 = x^2 + y^2 \,.$$

4.1 Properties

The modulus is a positive defined quantity

$$|z| \geq 0 \,,$$

where the equality holds if and only if $z = 0$, as can be seen easily from Eq. (4.0.1).

Moreover the modulus of the product of two complex numbers $z_1, z_2 \in \mathbb{C}$ is equal to the product of their moduli, in fact

$$|z_1 z_2| = \sqrt{\operatorname{Re}(z_1 z_2)^2 + \operatorname{Im}(z_1 z_2)^2}$$

and, from Eq. (2.3.1) and Eq. (2.3.2) we have

$$
\begin{aligned}
|z_1 z_2| &= \sqrt{(x_1 x_2 - y_1 y_2)^2 + (x_1 y_2 + x_2 y_1)^2} \\
&= \Big(x_1^2 x_2^2 + y_1^2 y_2^2 - 2x_1 x_2 y_1 y_2 \\
&+ x_1^2 y_2^2 + x_2^2 y_1^2 + 2x_1 x_2 y_1 y_2 \Big)^{1/2},
\end{aligned}
$$

from which

$$
\begin{aligned}
|z_1 z_2| &= \sqrt{x_2^2(x_1^2 + y_1^2) + y_2^2(x_1^2 + y_1^2)} \\
&= \sqrt{(x_1^2 + y_1^2)(x_2^2 + y_2^2)} \\
&= \sqrt{x_1^2 + y_1^2}\sqrt{x_2^2 + y_2^2} = |z_1||z_2|.
\end{aligned}
$$

4.2 Triangle inequality

Given two complex numbers $z_1, z_2 \in \mathbb{C}$ we have the so-called triangle inequality

$$\Big| |z_1| - |z_2| \Big| \le |z_1 + z_2| \le |z_1| + |z_2|.$$

For example, let's consider the second inequality, i.e.

$$|z_1 + z_2| \leq |z_1| + |z_2|,$$

that can be written also as

$$|(x_1 + x_2)^2 + (y_1 + y_2)^2| \leq \sqrt{x_1^2 + y_1^2} + \sqrt{x_2^2 + y_2^2},$$

where we have used the cartesian forms

$$z_1 = x_1 + iy_1, \qquad z_2 = x_2 + iy_2.$$

Squaring both sides of the previous inequality we obtain

$$(x_1 + x_2)^2 + (y_1 + y_2)^2 \leq \left(\sqrt{x_1^2 + y_1^2} + \sqrt{x_2^2 + y_2^2} \right)$$
$$\leq x_1^2 + y_1^2 + x_2^2 + y_2^2 + 2\sqrt{x_1^2 + y_1^2}\sqrt{x_2^2 + y_2^2},$$

or also, simplifying,

$$x_1 x_2 + y_1 y_2 \leq \sqrt{x_1^2 + y_1^2}\sqrt{x_2^2 + y_2^2}.$$

Squaring both sides again

$$(x_1 x_2 + y_1 y_2)^2 \leq (x_1^2 + y_1^2)(x_2^2 + y_2^2),$$

from which

$$x_1^2 x_2^2 + y_1^2 y_2^2 + 2x_1 x_2 y_1 y_2$$
$$\leq x_1^2 x_2^2 + x_2^2 y_1^2 + x_1^2 y_2^2 + y_1^2 y_2^2,$$

or

$$2x_1x_2y_1y_2 \leq x_2^2y_1^2 + x_1^2y_2^2 \,,$$

that can be written as

$$x_2^2y_1^2 + x_1^2y_2^2 - 2x_1x_2y_1y_2 \geq 0 \,,$$
$$(x_2y_1 - x_1y_2)^2 \geq 0 \,,$$

which is always true.

Chapter 5

Graphical and vector representations

The complex numbers are in biunivocal correspondence with the points of a plane where on the axes we report the real and imaginary parts. In this new picture the complex number $z = x + iy$ is related to the point $P(x, y)$, as shown in Figure 5.1.

The point P is in turn in biunivocal correspondence with the vector $\vec{OP}$, where the length of $\overline{OP}$, called ρ, is the length of the vector and the modulus of z

$$\rho = |z| = \sqrt{x^2 + y^2}\,.$$

Figure 5.1:

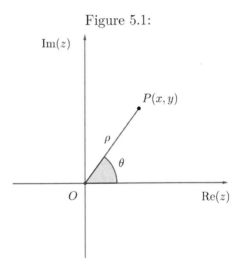

Chapter 6

Polar form

The polar form of a generic complex number $z = x + iy$ is

$$z = \rho(\cos\theta + i\sin\theta), \qquad (6.0.1)$$

where ρ and θ are the modulus and the argument of z, respectively, i.e.

$$\rho = \sqrt{x^2 + y^2}$$

and

$$\arg(z) = \theta.$$

As can be seen in Figure 5.1 and as said before, ρ is also the length of $\overline{OP}$, while θ is the angle that OP makes with the positive x axis.

Chapter 7

Exponential form

The exponential form of a generic complex number $z = x + iy$ is

$$z = \rho e^{i\theta},$$

that follows from the polar form together with the identity

$$e^{i\theta} = \cos\theta + i\sin\theta, \qquad (7.0.1)$$

called the Euler's formula.

The last expression can be demonstrated using the Taylor

series for the sine, the cosine and the exponential, in fact

$$
\begin{aligned}
e^{i\theta} &= \sum_{k=0}^{\infty} \frac{(i\theta)^k}{k!} = \sum_{\text{odd } k} \frac{(i\theta)^k}{k!} + \sum_{\text{even } k} \frac{(i\theta)^k}{k!} \\
&= \sum_{j=0}^{\infty} \frac{(i\theta)^{2j+1}}{(2j+1)!} + \sum_{j=0}^{\infty} \frac{(i\theta)^{2j}}{(2j)!} \\
&= \sum_{j=0}^{\infty} \frac{(i\theta)^{2j+1}}{(2j+1)!} + \sum_{j=0}^{\infty} \frac{(i\theta)^{2j}}{(2j)!}
\end{aligned}
$$

and, knowing that

$$
(i)^{2j+1} = i(i)^{2j} = i(i^2)^j = i(-1)^j , \qquad j \in \mathbb{N},
$$
$$
(i)^{2j} = (i^2)^j = (-1)^j , \qquad j \in \mathbb{N},
$$

we can write

$$
\begin{aligned}
e^{i\theta} &= \sum_{j=0}^{\infty} \frac{\theta^{2j+1} i (-1)^j}{(2j+1)!} + \sum_{j=0}^{\infty} \frac{\theta^{2j}(-1)^j}{(2j)!} \\
&= i\sin\theta + \cos\theta .
\end{aligned}
$$

From Eq. (7.0.1), in the case of $\theta = \pi$ we obtain the well known relation

$$
e^{i\pi} = \cos\pi + i\sin\pi = -1 .
$$

7.1 Euler's formula for trigonometry

Starting from Eq. (7.0.1) we can write

$$
e^{-i\theta} = \cos(-\theta) + i\sin(-\theta) = \cos\theta - i\sin\theta .
$$

By adding the latest expression to that showed in Eq. $(7.0.1)$ we obtain

$$
\begin{aligned}
e^{i\theta} + e^{-i\theta} &= \cos\theta + i\sin\theta + \cos\theta - i\sin\theta \\
&= 2\cos\theta \, ,
\end{aligned}
$$

from which

$$
\cos\theta = \frac{e^{i\theta} + e^{-i\theta}}{2} \, . \tag{7.1.1}
$$

Analogously we can write

$$
\begin{aligned}
e^{i\theta} - e^{-i\theta} &= \cos\theta + i\sin\theta - (\cos\theta - i\sin\theta) \\
&= \cos\theta + i\sin\theta - \cos\theta + i\sin\theta \\
&= 2i\sin\theta \, ,
\end{aligned}
$$

from which

$$
\sin\theta = \frac{e^{i\theta} - e^{-i\theta}}{2i} \, . \tag{7.1.2}
$$

The expressions showed in Eq. $(7.1.1)$ and Eq. $(7.1.2)$ are called the Euler's formula for trigonometry.

7.2 Phase factors

A complex number of the form

$$
e^{i\theta} \, ,
$$

where $\theta \in \mathbb{R}$, is called a phase factor, since it has modulus 1, as can be seen by calculating

$$|e^{i\theta}| = \sqrt{e^{i\theta}(e^{i\theta})^*} = \sqrt{e^{i\theta}e^{-i\theta}} = 1. \qquad (7.2.1)$$

7.3 Modulus of the exponential

We calculate now explicitly the modulus of the exponential.

Consider the exponential

$$e^z,$$

where $z \in \mathbb{C}$. We can write

$$
\begin{aligned}
|e^z| &= |e^{x+iy}| = |e^x||e^{iy}| = \left|e^{\mathrm{Re}(z)}\right||e^{iy}| \\
&= e^{\mathrm{Re}(z)}|e^{iy}|,
\end{aligned}
$$

and finally, being $y \in \mathbb{R}$, from Eq. (7.2.1) we have

$$|e^z| = e^{\mathrm{Re}(z)}. \qquad (7.3.1)$$

Chapter 8

De Moivre's theorem

The exponential form is particularly useful to calculate and interpret the product of two complex numbers. Consider two complex numbers z_1 and z_2 written in exponential form as

$$z_1 = \rho_1 e^{i\theta_1}, \qquad z_2 = \rho_2 e^{i\theta_2},$$

their product is

$$z_1 z_2 = \rho_1 e^{i\theta_1} \rho_2 e^{i\theta_2} = \rho_1 \rho_2 e^{i(\theta_1 + \theta_2)}, \qquad (8.0.1)$$

so that its modulus is given by the product of the two moduli $|z_1|$ and $|z_2|$ (as known) and its argument is given by the sum of the two arguments θ_1 and θ_2,

$$|z_1 z_2| = \rho_1 \rho_2 = |z_1||z_2|,$$
$$\arg(z_1 z_2) = \theta_1 + \theta_2 = \arg(z_1) + \arg(z_2).$$

We observe that the vector z_1 of the complex plane has

been dilated or contracted by a factor ρ_2 and rotated by an angle θ_2.

Recalling the polar form of Eq. (6.0.1), the Eq. (8.0.1) can be written also as

$$z_1 z_2 = \rho_1 \rho_2 \big(\cos(\theta_1 + \theta_2) + i \sin(\theta_1 + \theta_2) \big).$$

For the ratio z_1/z_2 we have similarly

$$\frac{z_1}{z_2} = \frac{\rho_1 e^{i\theta_1}}{\rho_2 e^{i\theta_2}} = \frac{\rho_1}{\rho_2} e^{i(\theta_1 - \theta_2)},$$

i.e.

$$\left| \frac{z_1}{z_2} \right| = \frac{\rho_1}{\rho_2} = \frac{|z_1|}{|z_2|},$$

$$\arg \left(\frac{z_1}{z_2} \right) = \theta_1 - \theta_2 = \arg(z_1) - \arg(z_2)$$

and, in polar form

$$\frac{z_1}{z_2} = \frac{\rho_1}{\rho_2} \big(\cos(\theta_1 - \theta_2) + i \sin(\theta_1 - \theta_2) \big).$$

Consider now the n-th power of a generic complex number

$$z = \rho e^{i\theta},$$

we can calculate

$$z^n = \left(\rho e^{i\theta} \right)^n = \rho^n e^{in\theta},$$

from which

$$|z^n| = \rho^n = |z|^n \,,$$

$$\arg(z^n) = n\theta = n\arg(z)$$

and

$$z^n = \rho^n\big(\cos(n\theta) + i\sin(n\theta)\big) \,,$$

where the last expression is called De Moivre's theorem. The generalization of this expression is the case of the product of $n \in \mathbb{N}$ complex numbers z_1, z_2, $\cdots$, z_n, we have

$$\begin{aligned} z_1 z_2 \cdots z_n &= \rho_1 \rho_2 \cdots \rho_n e^{i(\theta_1 + \theta_2 + \cdots + \theta_n)} \\ &= \left(\prod_{k=1}^{n} \rho_k\right) e^{i \sum_{k=1}^{n} \theta_k} \,. \end{aligned}$$

8.1 Roots

We want to calculate the n-th root of a complex number

$$z = \rho e^{i\theta} \,.$$

First of all we observe that the complex numbers

$$\rho e^{i(\theta + 2k\pi)} \,, \qquad k \in \mathbb{Z}$$

are all representations of the same number z, where k represents the number of turns around the origin.

There are functions that gives different results for different choices of k, the so-called multi-valued functions. The n-th root is one of them, in fact we can write

$$z^{1/n} = \left(\rho e^{i(\theta+2k\pi)}\right)^{1/n} = \rho^{1/n} e^{i(\theta+2k\pi)/n} \qquad (8.1.1)$$

and we have n distinct results, one for each k going from $k = 0$ to $k = n - 1$. In fact when $k = n$ we recover the same number of $k = 0$, since they differ by a complete turn around the origin.

The list of the n distinct n-th roots of z is

$$\rho^{1/n} e^{i\theta/n} , \ \ \rho^{1/n} e^{i(\theta+2\pi)/n} , \ \ \rho^{1/n} e^{i(\theta+4\pi)/n} , \ \cdots ,$$
$$\rho^{1/n} e^{i(\theta+2(n-2)\pi)/n} , \ \ \rho^{1/n} e^{i(\theta+2(n-1)\pi)/n} .$$

For example, concerning the square root of z we have two distinct roots, i.e.

$$\sqrt{\rho}\, e^{i\theta/2} , \qquad \sqrt{\rho}\, e^{i(\theta/2+\pi/2)} .$$

Graphically, the n n-th roots of $z = \rho e^{i\theta}$ are situated along the circumference centered in the origin of radius $\sqrt[n]{\rho}$, representing the vertices of a regular n-sides polygon inscribed.

8.2 Roots of unity

We can consider as a particular case the calculations of the n-th root of unity, namely the solutions of the equation

$$z^n = 1 \,.$$

Following Eq. (8.1.1) we can write simply

$$z = e^{i(\theta + 2k\pi)/n} \,,$$

or, in polar form, from Eq. (6.0.1),

$$z = \cos\left(\frac{\theta + 2k\pi}{/n}\right) + i \, \sin\left(\frac{\theta + 2k\pi}{/n}\right) \,,$$

being $\rho = |1| = 1$.

Chapter 9

Fundamental theorem of algebra

A very important theorem concerning the polynomial equations of the type

$$a_0 + a_1 z + a_2 z^2 + \cdots + a_{n-1} z^{n-1} + a_n z^n = 0 \,,$$

where a_k, for $k = 1, 2, \cdots, n$ are complex numbers with $a_n \neq 0$, called fundamental theorem of algebra, states that an equation of this form have exactly n complex solutions (some of them maybe repeated).

Chapter 10

Useful tips

- The complex numbers are not an ordered field, therefore you can't use the symbols $<$, $\leq$, $>$ or $\geq$ to order the complex numbers.

  These can be used with real numbers or, for example, with moduli of complex numbers.

- The modulus of a phase factor e^{ix}, $x \in \mathbb{R}$, is 1, i.e. $|e^{ix}| = 1$, for more details see Eq. (7.2.1).

- The square of the imaginary unit is -1, i.e. $i^2 = -1$, while its squared modulus is 1, $|i|^2 = 1$, as can be easily seen since the complex conjugate of i is $-i$,

$$(i)^* = -i$$

and

$$|i|^2 = i(i)^* = i(-i) = -i^2 = 1 \,.$$

- The reciprocal of the imaginary unit coincides with its conjugate complex, in fact from $i^2 = -1$ we can write

$$i = -\frac{1}{i}$$

and hence

$$\frac{1}{i} = -i = (i)^*.$$

Chapter 11

Exercises

11.1 Exercise 1

Calculate the following quantities:

- $(1 - 5i)(-2 + i)$;

- $\mathrm{Re}\,(i/(2 - 3i))$;

- $|z^3(z^* - i)|$, with $z = 1 - i$.

11.2 Exercise 2

Calculate all the 8-th roots of the unity.

11.3 Exercise 3

Calculate the 3-th roots of the complex number

$$z = \sqrt{3} + 3i\,.$$

11.4 Exercise 4

Solve the equation

$$z^2 + (1 + i)z + 3 - 2i = 0 .$$

11.5 Exercise 5

Calculate the quantity

$$\text{Im}(e^{iz}) ,$$

with $z = 2e^{i\pi/4}$.

11.6 Exercise 6

Calculate the following quantities:

- $\frac{z+z^*-i}{|z|^2}$, with $z = 3i + 1$;

- $\text{Im}(|3 + 4i|i - e^{i\pi/3})$;

- $\left| \frac{2i+1}{i-\sqrt{2}} \right|$, with $z = 1 - i$.

11.7 Exercise 7

Calculate the square roots of the complex number

$$z = 2 - i .$$

11.8 Exercise 8

Calculate the quantity

$$\left| e^{ikz} \right| ,$$

with $k \in \mathbb{R}$ and $z \in \mathbb{C}$.

Chapter 12

Solutions

12.1   Exercise 1

12.1.1   Text

Calculate the following quantities:

- $(1 - 5i)(-2 + i)$;

- $\mathrm{Re}\left(\frac{i}{2-3i}\right)$;

- $|z^3(z^* - i)|$, with $z = 1 - i$.

12.1.2   Solution

For the first point we can write

$$
\begin{aligned}
(1 - 5i)(-2 + i) &= -2 + 10i + i - 5i^2 \\
&= -2 + 11i + 5 \\
&= 3 + 11i .
\end{aligned}
$$

For the second point we have

$$
\begin{aligned}
\mathrm{Re}\left(\frac{i}{2-3i}\right) &= \mathrm{Re}\left(\frac{i}{2-3i}\cdot\frac{2+3i}{2+3i}\right) \\
&= \mathrm{Re}\left(\frac{i(2+3i)}{(2-3i)(2+3i)}\right) \\
&= \mathrm{Re}\left(\frac{2i-3}{2^2+3^2}\right) \\
&= \mathrm{Re}\left(\frac{2}{13}i-\frac{3}{13}\right) = -\frac{3}{13}\,.
\end{aligned}
$$

Finally, for the third point we calculate

$$
|z^3(z^*-i)| = |z^3||(z^*-i)| = |z|^3|(z^*-i)|\,,
$$

using $z = 1-i$ we obtain

$$
\begin{aligned}
|z^3(z^*-i)| &= |1-i|^3\big|\big((1-i)^*-i\big)\big| \\
&= \left(\sqrt{1^2+(-1)^2}\right)^3|(1+i-i)| \\
&= 2^{3/2}|1| = 2^{3/2}\,.
\end{aligned}
$$

12.2 Exercise 2

12.2.1 Text

Calculate all the 8-th roots of the unity.

12.2.2   Solution

The unity can be written in exponential form as

$$1 = e^{2k\pi i}, \qquad k \in \mathbb{Z},$$

therefore its 8-th roots are

$$1^{1/8} = e^{2k\pi i/8}, \qquad k \in \mathbb{Z},$$

or explicitly

$$k = 0 \ \rightarrow \ 1, \qquad k = 1 \ \rightarrow \ e^{(\pi/4)i},$$

$$k = 2 \ \rightarrow \ e^{(\pi/2)i}, \qquad k = 3 \ \rightarrow \ e^{(3\pi/4)i},$$

$$k = 4 \ \rightarrow \ e^{i\pi} = -1, \qquad k = 5 \ \rightarrow \ e^{(5\pi/4)i},$$

$$k = 6 \ \rightarrow \ e^{(3\pi/2)i}, \qquad k = 7 \ \rightarrow \ e^{(7\pi/4)i}.$$

12.3   Exercise 3

12.3.1   Text

Calculate the 3-th roots of the complex number

$$z = \sqrt{3} + 3i.$$

12.3.2 Solution

Firstly we write the complex number $z = \sqrt{3} + 3i$ in exponential form. The modulus is

$$\rho = |z| = \sqrt{(\sqrt{3})^2 + 3^2} = \sqrt{12} = 2\sqrt{3},$$

while the argument, $0 \leq \theta \leq 2\pi$, can be written as

$$\theta = \arctan \frac{3}{\sqrt{3}} = \arctan \sqrt{3} = \frac{\pi}{3}.$$

The number in exponential form is

$$z = \sqrt{3} + 3i = 2\sqrt{3}e^{i\pi/3}$$

and we can obtain the 3-th roots as

$$\begin{aligned} z^{1/3} &= \left(2\sqrt{3}e^{i(\pi/3+2k\pi)}\right)^{1/3} \\ &= \sqrt[6]{12}\,e^{i(\pi/9+2k\pi/3)}, \qquad k \in \mathbb{Z}. \end{aligned}$$

The 3 roots, for $k = 0, 1, 2$, are

$$k = 0 \ \rightarrow \ \sqrt[6]{12}\,e^{(\pi/9)i},$$
$$k = 1 \ \rightarrow \ \sqrt[6]{12}\,e^{i(\pi/9+2\pi/3)} = \sqrt[6]{12}\,e^{(7\pi/9)i},$$
$$k = 2 \ \rightarrow \ \sqrt[6]{12}\,e^{i(\pi/9+4\pi/3)} = \sqrt[6]{12}\,e^{(13\pi/9)i}.$$

12.4 Exercise 4

12.4.1 Text

Solve the equation

$$2z^2 + 3(i - 1)z + 6 - 4i = 0 \,.$$

12.4.2 Solution

We calculate the discriminant

$$
\begin{aligned}
\Delta &= 3^2(i - 1)^2 - 8(6 - 4i) \\
&= 9(-1 + 1 - 2i) - 48 + 32i \\
&= -18i - 48 + 32i \\
&= 48 + 14i = 2(24 + 7i) \,,
\end{aligned}
$$

from which

$$z_{1,2} = \frac{-3(i - 1) \pm \sqrt{24 + 7i}\sqrt{2}}{4} \,. \qquad (12.4.1)$$

We need to calculate the principal square root of $24 + 7i$. We write it in polar form

$$24 + 7i = 25\left(\cos\theta + i\sin\theta\right),$$

in fact

$$|24 + 7i| = \sqrt{24^2 + 7^2} = 25 \,,$$

where θ is its argument, with

$$\cos\theta = \frac{24}{25}, \qquad \sin\theta = \frac{7}{25}.$$

The principal square root is

$$\sqrt{24 + 7i} = \sqrt{25}\left(\cos(\theta/2) + i\sin(\theta/2)\right) \qquad (12.4.2)$$

and we can write

$$
\begin{aligned}
\cos(\theta/2) &= \pm\sqrt{(1 + \cos\theta)/2} \\
&= \pm\sqrt{(1 + 24/25)/2} \\
&= \pm\sqrt{49/50} = \pm\frac{7}{5\sqrt{2}} \\
&= \pm\frac{7\sqrt{2}}{10},
\end{aligned}
$$

but we know that the complex number $24 + 7i$ is represented by a point in the first quadrant, so that $0 < \theta < \pi/2$ and hence $0 < \theta/2 < \pi/4$ with $\cos(\theta/2) > 0$, therefore

$$\cos(\theta/2) = \frac{7\sqrt{2}}{10}.$$

Analogously

$$
\begin{aligned}
\sin(\theta/2) &= \pm\sqrt{(1 - \cos\theta)/2} \\
&= \pm\sqrt{(1 - 24/25)/2} \\
&= \pm\sqrt{1/50} = \pm\frac{1}{5\sqrt{2}} \\
&= \pm\frac{\sqrt{2}}{10},
\end{aligned}
$$

for the same reasons related to the choice of $\cos(\theta/2)$ we know that $\sin(\theta/2) > 0$ so that

$$
\sin(\theta/2) = \frac{\sqrt{2}}{10}.
$$

The expression of Eq. (12.4.2) becomes

$$
\begin{aligned}
\sqrt{24 + 7i} &= \sqrt{25}\left(\frac{7\sqrt{2}}{10} + i\frac{\sqrt{2}}{10}\right) \\
&= \frac{7\sqrt{50}}{10} + i\frac{\sqrt{50}}{10} \\
&= \frac{7\sqrt{2}}{2} + i\frac{\sqrt{2}}{2}.
\end{aligned}
$$

The solutions, from Eq. (12.4.1), become

$$
\begin{aligned}
z_{1,2} &= \frac{-3(i - 1) \pm \left(\frac{7\sqrt{2}}{2} + i\frac{\sqrt{2}}{2}\right)\sqrt{2}}{4} \\
&= \frac{-3i + 3 \pm (7 + i)}{4},
\end{aligned}
$$

the first solution is

$$
\begin{aligned}
z_1 &= \frac{-3i + 3 + (7 + i)}{4} \\
&= \frac{10 - 2i}{4} = \frac{5}{2} - \frac{1}{2}i
\end{aligned}
$$

and the second solution is

$$
\begin{aligned}
z_2 &= \frac{-3i + 3 - (7 + i)}{4} \\
&= \frac{-4 - 4i}{4} = -1 - i \, .
\end{aligned}
$$

12.5 Exercise 5

12.5.1 Text

Calculate the quantity

$$
\mathrm{Im}(e^{iz}) \, ,
$$

with $z = 2e^{i\pi/4}$.

12.5.2 Solution

It is convenient to write the given complex number z in cartesian form

$$
\begin{aligned}
z &= 2e^{i\pi/4} = 2\big(\cos(\pi/4) + i\sin(\pi/4) \big) \\
&= 2\left(\frac{\sqrt{2}}{2} + i\frac{\sqrt{2}}{2} \right) \\
&= \sqrt{2} + i\sqrt{2} \, .
\end{aligned}
$$

Calculating

$$
\begin{aligned}
\mathrm{Im}(e^{iz}) &= \mathrm{Im}\left(e^{i(\sqrt{2}+i\sqrt{2})}\right) \\
&= \mathrm{Im}\left(e^{i\sqrt{2}-\sqrt{2}}\right) \\
&= \mathrm{Im}\left(e^{i\sqrt{2}}e^{-\sqrt{2}}\right) ,
\end{aligned}
$$

from which, finally,

$$
\begin{aligned}
\mathrm{Im}(e^{iz}) &= \mathrm{Im}\left(e^{-\sqrt{2}}(\cos\sqrt{2}+i\sin\sqrt{2})\right) \\
&= \mathrm{Im}\left(e^{-\sqrt{2}}\cos\sqrt{2}+ie^{-\sqrt{2}}\sin\sqrt{2}\right) \\
&= e^{-\sqrt{2}}\sin\sqrt{2} = \frac{\sin\sqrt{2}}{e^{\sqrt{2}}} .
\end{aligned}
$$

12.6 Exercise 6

12.6.1 Text

Calculate the following quantities:

- $\frac{z+z^*-i}{|z|^2}$, with $z = 3i + 1$;

- $\mathrm{Im}(|3+4i|i - e^{i\pi/3})$;

- $\left|\frac{2i+1}{i-\sqrt{2}}\right|$, with $z = 1 - i$.

12.6.2 Solution

For the first point we calculate

$$
\frac{z+z^*-i}{|z|^2} = \frac{2\,\mathrm{Re}(z)-i}{|z|^2} ,
$$

being $z = 3i + 1$ we have

$$|z| = \sqrt{3^2 + 1^2} = \sqrt{10}, \qquad \mathrm{Re}(z) = 1,$$

so that

$$\frac{z + z^* - i}{|z|^2} = \frac{2 - i}{10} = \frac{1}{5} - i\,\frac{1}{10}\,.$$

For the second point we can write

$$\mathrm{Im}\big(|3 + 4i|i - e^{i\pi/3}\big)$$
$$= \mathrm{Im}\Big(\sqrt{3^2 + 4^2}\,i - \big(\cos(\pi/3) + i\sin(\pi/3)\big)\Big)$$
$$= \mathrm{Im}\big(5i - (1/2 + i\,\sqrt{3}/2)\big)$$
$$= \mathrm{Im}\big((5 - \sqrt{3}/2)i - 1/2\big) = 5 - \frac{\sqrt{3}}{2}\,.$$

Finally, for the third point,

$$\left|\frac{2i + 1}{i - \sqrt{2}}\right| = \frac{|2i + 1|}{|i - \sqrt{2}|} = \frac{\sqrt{2^2 + 1^2}}{\sqrt{1^2 + (-\sqrt{2})^2}}\,,$$

from which

$$\left|\frac{2i + 1}{i - \sqrt{2}}\right| = \frac{\sqrt{5}}{\sqrt{3}} = \sqrt{\frac{5}{3}}\,.$$

12.7 Exercise 7

12.7.1 Text

Calculate the square roots of the complex number

$$z = 2 - i \, .$$

12.7.2 Solution

Consider the polar form of z

$$z = \rho(\cos\theta + i\sin\theta) \, ,$$

with

$$\rho = |z| = |2 + i| = \sqrt{2^2 + 1^2} = \sqrt{5} \, .$$

The two square roots of z are

$$
\begin{aligned}
z^{1/2} &= \sqrt{\rho}\left(\cos\frac{\theta + 2k\pi}{2} + i\sin\frac{\theta + 2k\pi}{2}\right) \\
&= \sqrt[4]{5}\left(\cos\frac{\theta + 2k\pi}{2} + i\sin\frac{\theta + 2k\pi}{2}\right) \, ,
\end{aligned}
$$

with $k = 0, 1$, i.e.

$$
\begin{aligned}
k = 0 \ &\rightarrow \ \sqrt[4]{5}\big(\cos(\theta/2) + i\sin(\theta/2)\big) \, , \\
k = 1 \ &\rightarrow \ \sqrt[4]{5}\big(\cos(\theta/2) - i\sin(\theta/2)\big) \, .
\end{aligned}
\tag{12.7.3}
$$

We need to evaluate the quantities $\cos(\theta/2)$ and $\sin(\theta/2)$. We found that

$$2 - i = \sqrt{5}(\cos\theta + i\sin\theta)\,,$$

from which

$$\cos\theta = \frac{2}{\sqrt{5}} = \frac{2\sqrt{5}}{5}\,,$$

$$\sin\theta = -\frac{1}{\sqrt{5}} = -\frac{\sqrt{5}}{5}$$

and θ is an angle in the fourth quadrant. It follows that $\theta/2$ is an angle in the second quadrant.

We can calculate

$$
\begin{aligned}
\cos(\theta/2) &= \pm\sqrt{(1 + \cos\theta)/2} \\
&= \pm\sqrt{(1 + 2\sqrt{5}/5)/2} \\
&= \pm\sqrt{(5 + 2\sqrt{5})/10}
\end{aligned}
$$

and, being $\theta/2$ an angle in the second quadrant, we obtain

$$\cos(\theta/2) = -\sqrt{(5 + 2\sqrt{5})/10}\,.$$

Similarly, we can write

$$
\begin{aligned}
\sin(\theta/2) &= \pm\sqrt{(1 - \cos\theta)/2} \\
&= \pm\sqrt{(1 - 2\sqrt{5}/5)/2} \\
&= \pm\sqrt{(5 - 2\sqrt{5})/10}
\end{aligned}
$$

and therefore

$$\sin(\theta/2) = \sqrt{(5 - 2\sqrt{5})/10}\,.$$

The two roots in Eq. (12.7.3) become

$$k = 0 \;\rightarrow\; \sqrt[4]{5}\left(-\sqrt{\frac{5 + 2\sqrt{5}}{10}} + i\,\sqrt{\frac{5 - 2\sqrt{5}}{10}}\right),$$

$$k = 1 \;\rightarrow\; \sqrt[4]{5}\left(-\sqrt{\frac{5 + 2\sqrt{5}}{10}} - i\,\sqrt{\frac{5 - 2\sqrt{5}}{10}}\right).$$

12.8 Exercise 8

12.8.1 Text

Calculate the quantity

$$\left| e^{ikz} \right|,$$

with $k \in \mathbb{R}$ and $z \in \mathbb{C}$.

12.8.2 Solution

We write z in polar form

$$z = \rho(\cos\theta + i\sin\theta)$$

and evaluate the quantity

$$
\begin{aligned}
\left| e^{ikz} \right| &= \left| e^{ik\rho(\cos\theta + i\sin\theta)} \right| = \left| e^{ik\rho\cos\theta + i^2 k\rho\sin\theta} \right| \\
&= \left| e^{ik\rho\cos\theta} e^{-k\rho\sin\theta} \right| \\
&= \left| e^{ik\rho\cos\theta} \right| \left| e^{-k\rho\sin\theta} \right| = \left| e^{-k\rho\sin\theta} \right|
\end{aligned}
$$

and

$$
\left| e^{ikz} \right| = e^{-k\rho\sin\theta} \,,
$$

where we have used the fact that ρ, $\cos\theta$ and k are real quantities. This result agree with the formula shown in Eq. (7.3.1), since

$$
\mathrm{Re}(ikz) = -k\rho\sin\theta \,.
$$

Part II

Special relativity

Introduction

This book is dedicated to Einstein's special relativity. The main topics are: postulates of relativity, events and Minkowski space-time, Lorentz transformations, metrics in scalar products (with an analysis on the metric tensor in an Euclidean and Minkowski space), intervals and their classification, effects of relativity such as time dilation and length contraction, speed transformations, equation of motion in relativistic dynamics, relativistic Lagrangian, kinetic, mass and total energy and their non-relativistic limits, four-momentum conservation.

©2020 Dr. A. Mangoni   75   I ed. 9798675717668

Chapter 13

Notations

Generally we will use the Greek letters to indicate the indices that take values $0, 1, 2, 3$ and the Latin ones for the indices that assume values $1, 2, 3$. We will use the Minkowski metric tensor with signature $(+, -, -, -)$, i.e.

$$
\eta = \begin{pmatrix} 1 & 0 & 0 & 0 \\ 0 & -1 & 0 & 0 \\ 0 & 0 & -1 & 0 \\ 0 & 0 & 0 & -1 \end{pmatrix}
$$

and the Einstein convention that implies summation over two repeated indices, one contravariant and one covariant, such as:

$$
q^\mu p_\mu = \sum_{\mu=0}^{3} q^\mu p_\mu \,.
$$

Finally we will use the natural units, i.e.

$$
\hbar = c = 1 \,.
$$

Chapter 14

The basis of relativity

14.1 Postulates

The theory of special relativity is based on the following postulates:

- Principle of relativity. Same form for physical laws in any inertial reference frame;

- Constant speed of light. The speed of light in vacuum, called c, is constant in all inertial reference frames.

The value of the speed of light in vacuum is

$$c = 299792458 \text{ m/s} \simeq 3 \cdot 10^8 \text{ m/s},$$

or also

$$c \simeq 1.08 \cdot 10^9 \text{ km/h},$$

i.e., just over a billion kilometers per hour.

14.2 Event

Given a reference frame, we define an event something that happens in a certain time and in a certain position. An event is something that happens at a certain point in space-time that we identify in a Cartesian reference with the coordinates of the vector (with four components, called a four-vector)

$$x^{\mu} = \left(ct, x^1, x^2, x^3\right),$$

where c is the speed of light in vacuum (constant), t is the time and the three rightmost components stand for the classic x, y, z (the exponent numbers do not indicate an exponentiation). In this notation the indices (ranging from 0 to 3, one value for each space-time component) of vectors (it can also extended to matrices) are placed at the top (as in this case and does not mean an exponentiation!) or at the bottom. The two notations depend on the so-called metric (or metric tensor) of the theory as we will see later. In our case, the theory of special relativity, the metric tensor implies the following relationship between components with an index at the top (called contravariant components) or at the bottom (covariant components):

$$x_\mu = (ct, x_1, x_2, x_3) = \left(ct, -x^1, -x^2, -x^3\right),$$

where, in the passage between one notation and the other, only the spatial components (for which the index takes values from 1 to 3) change sign, while the temporal one (for which the index assumes only the value 0) remains the same.

- Convention 1: when the indices of vectors are Greek letters then they will assume the values 0,1,2,3, while for Latin letters indices they are meant to assume the values 1,2,3 ("spatial " i.e. they select only the spatial components of the four-vectors);

- Convention 2: "Einstein's convention on the sum of repeated indices ", when there are multiple components of vectors (matrices, tensors, four-vectors) multiplied with each other then a sum on the repeated indices is implied (sum that goes from 0 to 3 if the index is represented by a Greek letter and from 1 to 3 if that index is represented by a Latin letter, see the convention 1 above).

In summary, an event is defined by the vector (four-vector) of components

$$x^\mu = \left(ct, x^1, x^2, x^3\right),$$

with

$$\mu = 0, 1, 2, 3\,,$$

(in fact we used a Greek letter) and we have

$$\begin{cases} x^0 = x_0 = ct \\ x^i = -x_i \end{cases},$$

where we remember that i, being an index indicated with Latin letter, takes only values from 1 to 3.

14.3 Interval

Given an inertial reference frame and two coordinate events

$$x_A^\mu = (ct_A, x_A^1, x_A^2, x_A^3)$$

and

$$x_B^\mu = (ct_B, x_B^1, x_B^2, x_B^3)\,,$$

(where A and B are chosen to distinguish the two events and are not indexes) the interval (squared) between these two events is

$$\begin{aligned} \Delta s^2 &= c^2(t_B - t_A)^2 - (x_B^1 - x_A^1)^2 \\ &- (x_B^2 - x_A^2)^2 - (x_B^3 - x_A^3)^2 \end{aligned}$$

which can be written, using the Einstein convention, as

$$
\begin{aligned}
\Delta s^2 &= \sum_{\mu=0}^{3} (x_B - x_A)^\mu (x_B - x_A)_\mu \\
&= (x_B - x_A)^\mu (x_B - x_A)_\mu \\
&= c^2(t_B - t_A)^2 + (x_B - x_A)^i (x_B - x_A)_i \\
&= c^2(t_B - t_A)^2 - \sum_{i=1}^{3} \left[(x_B - x_A)^i \right]^2 .
\end{aligned}
$$

Using the following notation for the vector

$$
\vec{x} = \left(x^1, x^2, x^3 \right),
$$

we can write shortly

$$
x^\mu = (ct, \vec{x})
$$

and

$$
x^\mu = (ct, -\vec{x}),
$$

from which

$$
\Delta s^2 = c^2 \Delta t^2 - \Delta \vec{x}^2 ,
$$

with

$$
\Delta \vec{x}^2 = (\Delta \vec{x})^2 .
$$

The infinitesimal interval between two events is then written as

$$ds^2 = c^2 dt^2 - d\vec{x}^2 \,,$$

with

$$d\vec{x}^2 = (d\vec{x})^2 \,.$$

Starting from the postulates of special relativity, the interval between two events is the same in all inertial reference frames. Typically, when referring to coordinates of two inertial reference frames we will add a "prime" (i.e.: ') symbol to the right of the letter. Therefore

$$\Delta s^2 = \Delta s'^2 \,, \quad ds^2 = ds'^2 \,.$$

Observe that the squared interval can be also negative, which is impossible in Euclidean space.

The interval between two events is said:

- time-like if the squared interval is positive;

- space-like if the squared interval is negative;

- light-like if the squared interval is null.

From the invariance of the intervals between inertial reference frames it follows that if an interval between two events is of a type in a certain inertial reference frame then it will be of the same type in all the others. It should be observed that two simultaneous events (in the inertial reference frame where they are, in fact simultaneity is referred to the reference frame, Lorentz transformations show that if two events are simultaneous in one system they are not in other distinct frames) are separated by space-like intervals.

Chapter 15
Lorentz transformations

Lorentz transformations are obtained from the postulates of special relativity and from the invariance of the intervals between inertial reference frames. They replace Galileo's transformations (which are a limit, for low velocities, of Lorentz transformations) and are used to mathematically relate the coordinates of two events in two inertial reference frames. Suppose, for simplicity, that a second inertial reference frame (where we indicate all the components by adding the symbol "prime", as said before) moves at a constant speed v along the positive axis of the abscissas (remember that here the abscissas of a vectors are related to the component 1) with respect to a first inertial reference frame and that at instant 0 the origins and axes of the two systems coincide (and therefore the corresponding axes in the two systems are parallel).

In this case Lorentz's transformations are the following

$$\begin{cases} ct' = \frac{ct - \beta x^1}{\sqrt{1 - \beta^2}} \\ x'^1 = \frac{x^1 - \beta ct}{\sqrt{1 - \beta^2}} \\ x'^2 = x^2 \\ x'^3 = x^3 \end{cases},$$

where

$$\beta = v/c\,.$$

The reverse transformations,

$$\begin{cases} ct = \frac{ct' + \beta x'^1}{\sqrt{1 - \beta^2}} \\ x^1 = \frac{x'^1 + \beta ct'}{\sqrt{1 - \beta^2}} \\ x^2 = x'^2 \\ x^3 = x'^3 \end{cases},$$

can be obtained trivially by exchanging v with $-v$ and all the coordinates with the "prime" symbol (') with those without and vice versa.

We also define the Lorentz factor as

$$\gamma = \frac{1}{\sqrt{1 - \beta^2}} = \frac{c}{\sqrt{c^2 - v^2}} = \frac{1}{\sqrt{1 - \frac{v^2}{c^2}}}\,.$$

Observe how, in order for this expression to make sense, it must necessarily be $v < c$, i.e. no inertial reference frame can have a speed higher than that of light in a vacuum compared to another inertial reference frame. From this fact it follows, in particular, that the Lorentz factor is always greater than or equal to 1. These transformations connect the coordinates of an event seen by two distinct inertial reference frames. Note that in this case, where the motion between the two inertial reference frames is only along the abscissa axis, the other two spatial components do not change. Finally, note how time is no longer absolute as in Galilei's transformations. More generally, if the second frame moved with a generic velocity $\vec{v}$ the Lorentz transformations would be, using

$$\vec{x} = \left(x^1, x^2, x^3\right),$$

the following

$$\begin{cases} ct' = \gamma(ct - \vec{\beta} \cdot \vec{x}) \\ x'^1 = \vec{x} + \vec{\beta} \cdot \left[\frac{(\gamma-1)\vec{\beta}\cdot\vec{x}}{\beta^2} - \gamma ct\right] \end{cases}$$

with

$$\vec{\beta} = \vec{v}/c$$

and

$$\beta^2 = \vec{\beta} \cdot \vec{\beta}.$$

Chapter 16

Length contraction

The length contraction effect is part of the phenomenology of special relativity. Suppose we have two inertial reference frames, where the second moves at speed v with respect to the first, along the x axis, in the positive direction. Suppose also that all the corresponding axes are parallel and the origins coincide at the time $t = t' = 0$. Consider a bar of length L_0 lying along the x axis, being at rest in the second reference frame. Its length is given, in the system where it is at rest, by

$$L_0 = x'^1_B - x'^1_A \, ,$$

where we have indicated with A and B the two ends of the bar. We can calculate the length of the bar in the first frame, using the Lorentz transformations

$$\begin{cases} ct' = \frac{ct - \beta x^1}{\sqrt{1-\beta^2}} \\ x'^1 = \frac{x^1 - \beta ct}{\sqrt{1-\beta^2}} \\ x'^2 = x^2 \\ x'^3 = x^3 \end{cases},$$

where we remember that

$$\beta = v/c \, .$$

Using the Lorentz factor

$$\gamma = \frac{1}{\sqrt{1 - \frac{v^2}{c^2}}} \, ,$$

we obtain

$$x'^1_A = \gamma(x^1_A - \beta ct_A)$$

and

$$x'^1_B = \gamma(x^1_B - \beta ct_B) \, .$$

Now we set

$$t_A = t_B$$

in fact, in order to talk about the length of the bar in the first frame, it is necessary to measure its ends at the same time. Therefore the length in the first frame is

$$L = x_B^1 - x_A^1$$

and, using also the previous formula, we obtain

$$
\begin{aligned}
L_0 &= x_B'^1 - x_A'^1 = \gamma(x_B^1 - \beta ct_B) - \gamma(x_A^1 - \beta ct_A) \\
&= \gamma(x_B^1 - \beta ct_A) - \gamma(x_A^1 - \beta ct_A) \\
&= \gamma(x_B^1 - x_A^1) = \gamma L \,,
\end{aligned}
$$

therefore finally

$$L = L_0/\gamma \,.$$

Being the Lorentz factor always greater than or equal to 1, we obtain that in any other inertial reference frame the length of the bar will be less than in the system in which it is at rest. This is why this effect is called length contraction. The maximum length, the one measured in the frame in which it is at rest, is called proper length. A similar effect (but concerning a dilation instead of a contraction) occurs for the time intervals as we will see in the next chapter.

Consider as an example of application two inertial reference frames S and S', with the second traveling at speed v with respect to the first one, along the abscissa axis in the positive direction, so that all the corresponding axes are parallel and the origins coincide at the time $t = t' = 0$. Let's put a bar of length L_0 at rest in the system S', with an end in the origin O', belonging to the first quadrant, but inclined of an angle θ' with respect to the x' axis. As said before, the lengths of the projections of the bar along the x' and y' axes in the system S' (where it is at rest) can be written as

$$\begin{cases} L_{0y'} = L_0 \sin \theta' \\ L_{0x'} = L_0 \cos \theta' \end{cases} .$$

In the S system we know that the bar is moving with speed v along the abscissa axis and for this reason the component of the bar parallel to x will be contracted, as previously seen, by a Lorentz factor

$$\gamma = \frac{1}{\sqrt{1 - \frac{v^2}{c^2}}} ,$$

i.e.

$$L_x = L_{0x'}/\gamma ,$$

while the component orthogonal to v does not suffer any contraction

$$L_y = L_{0y'} .$$

Still in the S system, the lengths of the projections of the bar on the axes are

$$\begin{cases} L_y = L\sin\theta \\ L_x = L\cos\theta \end{cases},$$

where L is the length of the bar and θ is the angle between it and the x axis, always in the S system.

Therefore starting from

$$\begin{cases} L_y = L\sin\theta = L_{0y'} \\ L_x = L\cos\theta = L_{0x'}/\gamma \end{cases},$$

we obtain

$$\begin{cases} L\sin\theta = L_0\sin\theta' \\ L\cos\theta = L_0\cos\theta'/\gamma \end{cases},$$

from which, for example,

$$\tan\theta = \gamma\tan\theta',$$

hence the angle between the bar and the abscissa axis is different in the two systems S and S'. In addition, we calculate the length of the bar using the Pythagorean theorem

$$L = \sqrt{L_x^2 + L_y^2}$$

and

$$L_0 = \sqrt{L_{0x'}^2 + L_{0y'}^2}\,,$$

hence

$$\begin{aligned}
L_0 &= \sqrt{\gamma^2 L_x^2 + L_y^2} = \sqrt{\gamma^2 L_x^2 + L_y^2 + L_x^2 - L_x^2} \\
&= \sqrt{L^2 + (\gamma^2 - 1)L_x^2} = L\sqrt{1 + (\gamma^2 - 1)\frac{L_x^2}{L^2}}\,.
\end{aligned}$$

Chapter 17

Time dilation

The effect of time dilation is part of the phenomenology of special relativity. Suppose, as in the previous chapter, to have two inertial reference frames, in which the second travels at speed v with respect to the first, along the x axis, in the positive direction, and such that all the corresponding axes are parallel and the origins coincide at the time $t = t' = 0$. Let's consider, this time, an event that occurs in the second reference frame, in a certain fixed spatial point and occurs between two instants of time whose difference is

$$\Delta t' = t'^1_B - t'^1_A.$$

To calculate the time interval for the same event in the first reference frame we can use the Lorentz transformations

$$\begin{cases} ct = \frac{ct' + \beta x'^1}{\sqrt{1 - \beta^2}} \\ x^1 = \frac{x'^1 + \beta ct'}{\sqrt{1 - \beta^2}} \\ x^2 = x'^2 \\ x^3 = x'^3 \end{cases},$$

with

$$\gamma = \frac{1}{\sqrt{1 - \frac{v^2}{c^2}}}.$$

Since

$$\beta = v/c,$$

we can write

$$t_A = \gamma(t'_A + \beta c x'^1_A/c)$$

and

$$t_B = \gamma(t'_B + \beta c x'^1_B/c),$$

hence the time interval in the first inertial reference frame

$$\Delta t = t_B - t_A = \gamma\left(t'_B - t'_A + \beta/c(x'^1_B - x'^1_A)\right),$$

 I ed. 9798675717668

i.e.

$$\Delta t = \gamma\left(\Delta t' + \beta(x_B'^1 - x_A'^1)/c\right).$$

By hypothesis we have

$$x_B'^1 = x_A'^1,$$

because the event in the second reference frame occurs at a fixed point. So finally

$$\Delta t = \gamma\Delta t'.$$

The time interval in the inertial reference frame where the event occurs at a fixed point is called the proper time interval and we observe that it is the smallest time interval for that event. In fact in any other inertial reference frame the same event will have a duration longer, as shown by the latter formula, being the Lorentz factor always greater than or equal to 1.

Chapter 18
The metric tensor

The scalar product between a four-vector x and itself, in the Minkowski space (space-time where events in special relativity occur, i.e. the Euclidean space $\mathbb{R}^3$ plus the time axis), can be defined as

$$x^2 = x^\mu x_\mu = (x^0)^2 - (x^i)^2 = \eta_{\mu\nu} x^\mu x^\nu \,,$$

where a sum is implied on the two final Greek indices (Einstein convention) and where the matrix called "metric tensor" or simply "metric" has the components

$$\eta_{\mu\nu} = \begin{pmatrix} 1 & 0 & 0 & 0 \\ 0 & -1 & 0 & 0 \\ 0 & 0 & -1 & 0 \\ 0 & 0 & 0 & -1 \end{pmatrix}.$$

Note that the implied summation eliminates all products (which would be sixteen like the terms of the η matrix) between components with different indices and leaves only the terms with components with the same index, because the elements out of the diagonal of the matrix are null. From the above formula we can write (remember the implied sum on the repeated index)

$$x_\mu = \eta_{\mu\nu} x^\nu$$

i.e., making the calculation (sum implied on alpha that goes from 0 to 3, being a Greek letter),

$$\eta_{\mu\alpha} \eta^{\alpha\nu} = \delta_\mu^\nu,$$

where we have used the so-called Kronecker delta, defined as

$$\delta_\mu^\nu = \delta_{\mu\nu} = \begin{cases} 0 & \text{if } \mu \neq \nu \\ 1 & \text{if } \mu = \nu \end{cases}.$$

It follows also

$$\eta^{\mu\nu} = \eta_{\mu\nu}$$

and it is therefore said that the metric tensor "raises or lowers" the indices since, summing up:

$$x_\mu = \eta_{\mu\nu} x^\nu$$

and

$$x^\mu = \eta^{\mu\nu} x_\nu \,.$$

The scalar product between any two four-vectors is then written, generalizing,

$$a \cdot b = a^\mu b^\nu \eta_{\mu\nu} = a^\mu b_\mu = a_\mu b^\mu \,.$$

In the three-dimensional Euclidean space, given two vectors of components

$$\vec{x} = (x^1, x^2, x^3)$$

and

$$\vec{y} = (y^1, y^2, y^3) \,,$$

their scalar product is given by

$$\vec{x} \cdot \vec{y} = \eta_{ij}^E x^i y^j = x^1 y^1 + x^2 y^2 + x^3 y^3$$

and therefore the Euclidean metric tensor can be written simply as

$$\eta_{ij}^E = \begin{pmatrix} 1 & 0 & 0 \\ 0 & 1 & 0 \\ 0 & 0 & 1 \end{pmatrix} \,.$$

This is the reason why using the Euclidean space it is not necessary to introduce the contravariant and covariant components for the vectors (high and low indices), because in this space they coincide, in fact the metric tensor is represented by the identity matrix.

Chapter 19
Transformation of velocities

Suppose to have two inertial reference frames, where the second travels at speed v with respect to the first along the x axis in the positive direction, and such that all the corresponding axes are parallel and the origins coincide at the time $t = t' = 0$. Consider the motion of a particle, along the abscissa axis, seen from the two reference frames. In the first one the particle has a velocity u given by

$$u = \frac{dx^1}{dt} \, ,$$

while in the second frame it has a velocity u' given by

$$u' = \frac{dx'^1}{dt'} \, .$$

We calculate, using the Lorentz transformations, these two quantities and try to find a relation between them.

We recall the Lorentz transformations and the inverse ones, which in this case, as in the others, are

$$\begin{cases} ct = \frac{ct+\beta x'^1}{\sqrt{1-\beta^2}} \\ x^1 = \frac{x'^1+\beta ct'}{\sqrt{1-\beta^2}} \\ x^2 = x^2 \\ x^3 = x^3 \end{cases}$$

and

$$\begin{cases} ct' = \frac{ct-\beta x^1}{\sqrt{1-\beta^2}} \\ x'^1 = \frac{x^1-\beta ct}{\sqrt{1-\beta^2}} \\ x'^2 = x^2 \\ x'^3 = x^3 \end{cases} ,$$

we also remember that

$$\beta = v/c$$

and

$$\gamma = \frac{1}{\sqrt{1 - \frac{v^2}{c^2}}} .$$

We have, calculating,

$$\frac{dx'^1}{dt'} = \frac{c\gamma(dx^1 - \beta cdt)}{\gamma(cdt - \beta dx^1)} = \frac{dx^1 - \beta cdt}{dt\left[1 - \beta dx^1/(cdt)\right]} ,$$

in fact the Lorentz factor and β are constant, hence

$$\frac{dx'^1}{dt'} = \frac{dx^1 - \beta c dt}{dt(1 - \beta u/c)} ,$$

finally

$$u' = \frac{u - v}{1 - \frac{vu}{c^2}} .$$

Chapter 20

Equation of motion

The Newton's law on dynamics, classically, is

$$\vec{F} = \frac{d\vec{p}}{dt} \, ,$$

where

$$\vec{p} = m\vec{v}$$

is the classical momentum of a particle with mass m and velocity v. The first formula is valid also in the framework of special relativity, but the momentum of a relativistic particle becomes

$$\vec{p} = m\gamma\vec{v} \, ,$$

with

$$\gamma = \frac{1}{\sqrt{1 - \frac{v^2}{c^2}}} \, .$$

This can be derived starting by the Lagrangian of a free particle that in classical non-relativistic mechanics can be written as

$$L = \frac{1}{2} m \dot{x}^2 \,,$$

where the point above the letter indicates the derivative with respect to time, while in relativistic mechanics has the form

$$L = -mc^2 \sqrt{1 - \frac{\dot{x}^2}{c^2}} \,.$$

The equations of motion imply a momentum given by the formula shown above. Indeed from

$$L = -mc^2 \sqrt{1 - \frac{\sum_i (\dot{x}^i)^2}{c^2}} \,,$$

we have

$$\frac{\partial L}{\partial x^i} = 0$$

and

$$\frac{\partial L}{\partial \dot{x}^i} = -mc^2 \frac{1}{2} \frac{-2\dot{x}^i / c^2}{\sqrt{1 - \dot{x}^2 / c^2}} \,,$$

therefore

$$\frac{\partial L}{\partial \dot{x}^i} = m\gamma \dot{x}^i \,.$$

From the Euler-Lagrange equations

$$\frac{d}{dt}\frac{\partial L}{\partial \dot{x}^i} = \frac{\partial L}{\partial x^i}, \quad i = 1, 2, 3,$$

i.e.

$$\vec{F} = \frac{d\vec{p}}{dt},$$

with

$$\vec{p} = m\gamma\vec{v}$$

and with

$$\vec{F} = 0,$$

that is correct, in fact we started from the Lagrangian for a free particle. The equation of motion is, in summary,

$$\vec{F} = \frac{d}{dt}(m\gamma\vec{v}) = m\gamma\frac{d\vec{v}}{dt} + m\vec{v}\frac{d\gamma}{dt},$$

where we assumed, in the last step, the independence of the mass by the time. The derivative showed in the last term is

$$\frac{d\gamma}{dt} = \frac{d}{dt}\left(1 - \frac{v^2}{c^2}\right)^{1/2} = -\frac{1}{2}\left(1 - \frac{v^2}{c^2}\right)^{-3/2}$$
$$\cdot \left(\frac{-2\vec{v}}{c^2}\right)\frac{d\vec{v}}{dt} = \frac{\gamma 3}{c^2}\vec{v} \cdot \frac{d\vec{v}}{dt}.$$

We observe that we can introduce a four-velocity defined as

$$u^{\mu} = \frac{dx^{\mu}}{d\tau} \,,$$

where τ is the proper time, related to the time t by

$$dt = \gamma \, d\tau \,,$$

with, we remember,

$$\gamma = \frac{1}{\sqrt{1 - \frac{v^2}{c^2}}} \,.$$

The four-velocity can also be written as

$$u^{\mu} = \gamma \frac{dx^{\mu}}{dt} = (\gamma c, \gamma \vec{v}) \,,$$

where

$$\vec{v} = \left(\frac{dx}{dt}, \frac{dy}{dt}, \frac{dz}{dt} \right)$$

is the velocity of the particle in a certain reference frame S.

In this way the four-momentum for a particle of a non-negligible mass m can be written as

$$p^{\mu} = mu^{\mu} \,.$$

Calculating its square we obtain

$$p^\mu p_\mu = m^2 u^\mu u_\mu \,,$$

with

$$u^\mu u_\mu = \gamma^2 c^2 - \gamma^2 \vec{v}^2 = \gamma^2 (c^2 - \vec{v}^2) \,,$$

from which, simplifying,

$$u^\mu u_\mu = \gamma^2 \frac{c^2}{\gamma^2} = c^2$$

and therefore

$$p^\mu p_\mu = m^2 c^2 \,.$$

Chapter 21
Total relativistic energy

The total energy of a relativistic particle of momentum p and mass m satisfies the relation

$$E^2 = m^2 c^4 + p^2 c^2$$

and we can write also

$$E = T + mc^2 \,,$$

where T is the kinetic energy and the second addend is the mass energy. The kinetic energy is

$$T = mc^2 (\gamma - 1)$$

and therefore the total energy becomes

$$E = mc^2 \gamma \,.$$

Note that for massless particles, only the first formula can be used, from which

$$E_{m=0} = |\vec{p}|c = pc \,.$$

Momentum and energy can be seen as components of a four-vector, the so-called four-momentum. It has contravariant coordinates given by

$$p^{\mu} = \left(\frac{E}{c}, \vec{p} \right)$$

and it transforms, by change of inertial reference frame, under the effect of Lorentz transformations in similar way as the position four-vector x.

Starting from the expression for the total energy of a free relativistic particle, we can find its non-relativistic limit, obtained when

$$\frac{v}{c} \to 0 \,,$$

i.e. when the velocities are much lower than that of light in the vacuum. To obtain the non-relativistic limit we calculate the series expansion of the Lorentz factor

$$
\begin{aligned}
\gamma(v) \quad \sim \quad & \gamma(0) + v \left(\frac{v\gamma^3}{c^2} \Big|_{v=0} \right) \\
+ \quad & \frac{1}{2}v^2 \left(\frac{\gamma^3 + 3v^2\gamma^5/c^2}{c^2} \Big|_{v=0} \right) \,,
\end{aligned}
$$

where we have used the following derivatives of the Lorentz factor

$$\gamma = \frac{1}{\sqrt{1 - \frac{v^2}{c^2}}} = \left(1 - \frac{v^2}{c^2}\right)^{-1/2},$$

i.e.

$$\frac{d\gamma}{dv} = -\frac{1}{2}\left(1 - \frac{v^2}{c^2}\right)^{-3/2}\left(-2\frac{v}{c^2}\right) = \frac{d\gamma}{dv} = \frac{v\gamma^3}{c^2}$$

and

$$
\begin{aligned}
\frac{d^2\gamma}{dv^2} &= \frac{d}{dv}\frac{v\gamma^3}{c^2} = \frac{\gamma^3 + 3v\gamma^2 d\gamma/dv}{c^2} \\
&= \frac{\gamma^3 + 3v^2\gamma^5/c^2}{c^2}.
\end{aligned}
$$

We obtain, finally,

$$\gamma \sim 1 + \frac{1}{2}\frac{v^2}{c^2}.$$

We replace this result in the expression of total energy

$$E = mc^2\gamma,$$

obtaining

$$E = mc^2\left(1 + \frac{1}{2}\frac{v^2}{c^2}\right) = mc^2 + \frac{1}{2}mv^2,$$

i.e. the sum of the energy at rest (not present in the non-relativistic theories) and the well-known expression for the non-relativistic kinetic energy.

Chapter 22

Conservation laws

In non-relativistic mechanics we know that for a free system, i.e. not subject to forces, the total momentum and the energy are conserved. In special relativity these two conservation laws become only one conservation law: the four-momentum

$$p^\mu = \left(\frac{E}{c}, \vec{p} \right)$$

conservation law.

The conservation, for a certain physical process, such as a decay of a particle, implies that the four-momentum before and after the process is the same

$$p^\mu = P^\mu \,,$$

where we have indicated with capital letter the total final four-momentum, while with lowercase the initial one.

Let us consider, as an example of application, the decay

of a particle X into two particles Y and Z. We can write the decay as

$$X(p) \to Y(k) + Z(q)\,,$$

where in parentheses are indicated the four-momenta of the particles. The conservation law of the four-momentum can be written as

$$p^{\mu} = k^{\mu} + q^{\mu}$$

and the final four-momentum is

$$P^{\mu} = k^{\mu} + q^{\mu}\,.$$

This law is valid in any inertial reference frame. To simplify calculations we take the square of both members

$$p^2 = p^{\mu}p_{\mu} = (k^{\mu} + q^{\mu})(k_{\mu} + q_{\mu}) = (k+q)^2\,,$$

namely

$$p^2 = k^2 + q^2 + 2k \cdot q\,.$$

Remembering that

$$p^2 = p^{\mu}p_{\mu} = m_X^2 c^2\,,$$
$$k^2 = k^{\mu}k_{\mu} = m_Y^2 c^2$$

and

$$q^2 = q^\mu q_\mu = m_Z^2 c^2\,,$$

where m_X, m_Y and m_Z are the masses of the three particles: X, Y and Z. The conservation of the four-momentum implies

$$m_X^2 c^2 = m_Y^2 c^2 + m_Z^2 c^2 + 2k \cdot q\,,$$

i.e.

$$k \cdot q = \frac{(m_X^2 - m_Y^2 - m_Z^2)c^2}{2}\,,$$

where the scalar product at first member (which is a Lorentz scalar, i.e. has the same value in each inertial reference frame) can be calculated in any reference frame, such as the center of mass (CM) system, where the decaying particle X is at rest.

Starting from the conservation of the four-momentum it is possible to derive also other relations, involving scalar products between different four-momenta, in fact from

$$p^\mu = k^\mu + q^\mu\,,$$

it follows that

$$k^\mu = p^\mu - q^\mu \,,$$
$$q^\mu = p^\mu - k^\mu \,,$$

and we can write

$$k^2 = p^2 + q^2 - 2p \cdot q \,,$$
$$q^2 = p^2 + k^2 - 2p \cdot k \,,$$

from which

$$p \cdot q = \frac{(m_X^2 - m_Y^2 + m_Z^2)c^2}{2} \,,$$
$$p \cdot k = \frac{(m_X^2 + m_Y^2 - m_Z^2)c^2}{2} \,.$$

In the CM system the three four-momenta have the form

$$p^\mu = (E_X/c, \vec{0}) \,,$$
$$k^\mu = (E_Y/c, \vec{k})$$

and

$$q^\mu = (E_Z/c, \vec{q}) \,,$$

being the X particle at rest. The energies and momenta of the particles are referred to the CM system and are not Lorentz invariant quantities, they can assume different values in different inertial reference frames. In the CM system the particles in the final state have opposite momenta, so they belong to the same straight line and we can write

$$\vec{k} = -\vec{q}$$

and also

$$|\vec{k}| = |\vec{q}| \,.$$

The four-momenta become

$$k^\mu = (E_Y/c, \vec{k}) \,,$$
$$q^\mu = (E_Z/c, -\vec{k}) \,.$$

Similarly, for the energy conservation law (temporal component of the four-momentum), we have

$$\frac{E_X}{c} = \frac{E_Y}{c} + \frac{E_Z}{c} \,,$$

namely

$$E_X = E_Y + E_Z \,,$$

the energy of the X particle in the CM system is its mass energy, being at rest, i.e.

$$E_X = m_X c^2 \,,$$

from which

$$m_X c^2 = E_Y + E_Z$$

and the three four-momenta, in the CM system, take the form

$$p^\mu = (m_X c, \vec{0}),$$
$$k^\mu = (E_Y/c, \vec{k})$$

and

$$q^\mu = (E_Z/c, -\vec{k}).$$

The scalar products calculated before, i.e.

$$k \cdot q = \frac{(m_X^2 - m_Y^2 - m_Z^2)c^2}{2},$$
$$p \cdot q = \frac{(m_X^2 - m_Y^2 + m_Z^2)c^2}{2},$$
$$p \cdot k = \frac{(m_X^2 + m_Y^2 - m_Z^2)c^2}{2},$$

in the CM system are written as

$$k \cdot q = \frac{E_Y E_Z}{c^2} + |\vec{k}|^2,$$
$$p \cdot k = m_X E_Y,$$
$$p \cdot q = m_X E_Z.$$

In particular, the latter two expressions, together with the previous ones, become

$$m_X E_Y = \frac{(m_X^2 + m_Y^2 - m_Z^2)c^2}{2}$$

and

$$m_X E_Z = \frac{(m_X^2 - m_Y^2 + m_Z^2)c^2}{2} \, ,$$

hence the energies in the CM system are

$$E_Y = \frac{(m_X^2 + m_Y^2 - m_Z^2)c^2}{2m_X}$$

and

$$E_Z = \frac{(m_X^2 - m_Y^2 + m_Z^2)c^2}{2m_X} \, .$$

Similarly, from the equation

$$\frac{E_Y E_Z}{c^2} + |\vec{k}|^2 = \frac{(m_X^2 - m_Y^2 - m_Z^2)c^2}{2}$$

we obtain

$$
\begin{aligned}
|\vec{k}|^2 &= \frac{(m_X^2 - m_Y^2 - m_Z^2)c^2}{2} - \frac{E_Y E_Z}{c^2} \\
&= \frac{(m_X^2 - m_Y^2 - m_Z^2)c^2}{2} \\
&\quad - \frac{(m_X^2 + m_Y^2 - m_Z^2)(m_X^2 - m_Y^2 + m_Z^2)c^2}{4m_X^2} \\
&= \frac{[m_X^4 + (m_Y^2 - m_Z^2)^2 - 2m_X^2(m_Y^2 + m_Z^2)]c^2}{4m_X^2} \\
&= \frac{\sqrt{m_X^4 + (m_Y^2 - m_Z^2)^2 - 2m_X^2(m_Y^2 + m_Z^2)}\, c}{2m_X}.
\end{aligned}
$$

The total energy is written as the sum of kinetic energy T and mass energy

$$
\begin{aligned}
E_Y = T_Y + m_Y c^2 &= \frac{(m_X^2 + m_Y^2 - m_Z^2)c^2}{2m_X}, \\
E_Z = T_Z + m_Z c^2 &= \frac{(m_X^2 - m_Y^2 + m_Z^2)c^2}{2m_X},
\end{aligned}
$$

from which

$$
\begin{aligned}
T_Y &= \frac{(m_X^2 + m_Y^2 - m_Z^2)c^2}{2m_X} - m_Y c^2 \\
&= \frac{(m_X^2 + m_Y^2 - m_Z^2 - 2m_X m_Y)c^2}{2m_X}
\end{aligned}
$$

and

$$
\begin{aligned}
T_Z &= \frac{(m_X^2 - m_Y^2 + m_Z^2)c^2}{2m_X} - m_Z c^2 \\
&= \frac{(m_X^2 - m_Y^2 + m_Z^2 - 2m_X m_Z)c^2}{2m_X}.
\end{aligned}
$$

Finally

$$T_Y = \frac{[(m_X - m_Y)^2 - m_Z^2]c^2}{2m_X}$$

and

$$T_Z = \frac{[(m_X - m_Z)^2 - m_Y^2]c^2}{2m_X} \, .$$

Part III

The mathematics of quantum mechanics

Introduction

In this book we expose the mathematics of quantum mechanics. The main topics are: vectors, ket and bra space, properties and operations, product for a scalar, internal product between ket and bra, norm and Schwarz inequality, orthogonality, operators and their operations, operator acting on kets as a measure of an observable for a physical state, adjoint operator, hermitian operators, unitary operator, external product, projectors, basis of eigenkets, representation of vectors and operators, matrix algebra.

Chapter 23

Ket space

In quantum mechanics we have to deal with physical states. A physical state is represented by a vector in an appropriate vectorial space with a certain dimension. This state vector is called ket, a term introduced by Dirac, and it contains all the information on the physical state. The dimension of the space depends on the situation, it can be finite or infinite and in the latter case it is called Hilbert space. A ket for a state a is indicated by the notation:

$$|a\rangle$$

and, in general, the vectorial space for kets is denoted by the letter V.

23.1 Sum of kets

The sum of two kets of the vectorial space V is still a ket belonging to the same space V, we write:

$$\forall \, |a\rangle, |b\rangle \in V \implies |a\rangle + |b\rangle \in V \, .$$

The sum of ket has the following properties:

Commutative property

Given two kets $|a\rangle$ and $|b\rangle$ we have the commutative property

$$|a\rangle + |b\rangle = |b\rangle + |a\rangle \, .$$

Associative property

Given three kets $|a\rangle$, $|b\rangle$ and $|c\rangle$ we have the associative property

$$\left(|a\rangle + |b\rangle\right) + |c\rangle = |a\rangle + \left(|b\rangle + |c\rangle\right) \, .$$

23.2 Product for a scalar

The product of a ket of the vectorial space V with a complex number is still a ket belonging to the same space

V,

$$\forall \, |a\rangle \in V, c \in \mathbb{C} \implies c \cdot |a\rangle = |a\rangle \cdot c \in V \, .$$

Usually to indicate the product $c \cdot |a\rangle$ we use the simplest notation $c|a\rangle$. In particular, the product of a ket for the number zero provides the null ket, in fact we can introduce the null element for the vectorial space (the null ket, indicated with $|0\rangle$), such that

$$\forall \, |a\rangle \in V \implies 0|a\rangle = |0\rangle \in V \, ,$$
$$\forall \, |a\rangle \in V \implies |a\rangle + |0\rangle = |a\rangle \, .$$

The ket product has the following properties:

Associative property

Given a ket $|a\rangle$ and two complex numbers $c_1, c_2 \in \mathbb{C}$ we have the associative property

$$c_1 \cdot c_2 |a\rangle = c_1 \cdot (c_2 |a\rangle) = (c_1 c_2) |a\rangle \, .$$

Distributive property with respect to scalars

Given a ket $|a\rangle$ and two complex numbers $c_1, c_2 \in \mathbb{C}$ we have the distributive property with respect to scalars

$$(c_1 + c_2)|a\rangle = c_1 |a\rangle + c_2 |a\rangle \, .$$

Distributive property with respect to kets

Given two kets $|a\rangle$, $|b\rangle$ and a complex number $c \in \mathbb{C}$ we have the distributive property with respect to kets

$$c\Big(|a\rangle + |b\rangle\Big) = c|a\rangle + c|b\rangle \,.$$

23.3 Opposite of a ket

The opposite of ket $|a\rangle$ is the ket $|b\rangle$ such that their sum gives the null ket

$$|a\rangle + |b\rangle = |0\rangle \,.$$

In particular, each ket of a vectorial space V admits its opposite ket which satisfies the previous relation, i.e.

$$\forall\, |a\rangle \in V,\; \exists\, |b\rangle \in V : |a\rangle + |b\rangle = |0\rangle \,.$$

We indicate the opposite of ket $|a\rangle$ with

$$-|a\rangle \,,$$

from which

$$|a\rangle - |a\rangle = |0\rangle \,.$$

23.4   Difference of kets

The difference of two kets $|a\rangle$ and $|b\rangle$ of the vectorial space V is still a ket belonging to the same space V, with

$$\forall\, |a\rangle, |b\rangle \in V \implies |a\rangle - |b\rangle \in V\,,$$

where the difference is defined as

$$|a\rangle - |b\rangle = |a\rangle + (-|b\rangle) = |a\rangle + (-1)|b\rangle\,,$$

obtained by adding the opposite of the ket $|b\rangle$ to the ket $|a\rangle$. We observe that the commutative property does not generally apply to the difference

$$|a\rangle - |b\rangle \neq |b\rangle - |a\rangle\,.$$

23.5   Linear independence

Given n kets belonging to a vectorial space V, indicated with

$$|a_1\rangle, |a_2\rangle, \cdots, |a_n\rangle\,,$$

we say that they are linearly independent if the equation

$$c_1|a_1\rangle + c_2|a_2\rangle + \cdots + c_n|a_n\rangle = 0\,,$$

or, in compact form,

$$\sum_{k=1}^{n} c_k |a_k\rangle = 0$$

with

$$c_1, c_2, \cdots, c_n \in \mathbb{C},$$

has the trivial solution as its only solution, that is

$$c_k = 0, \quad \forall k.$$

Otherwise it will be said that the n kets are linearly dependent.

23.6   Postulate on physical states

We postulate that if $|a\rangle$ is the ket that represents a physical state so any other ket of the type $c|a\rangle$ with $c \in \mathbb{C}$ represents the same physical state.

In general, for a physical state, the multiplicative constant of the ket is chosen such that the resulting state vector is normalized to 1, similarly to the case of the wave function $(\psi(x))$ normalization for a particle, from which

$$\int \|\psi(x)\|^2 \, dx = 1.$$

This, applied to kets, requires the concepts of bra vectors and of the internal product between kets and bras, that will be discussed later.

Chapter 24

Bra space

In addition to the ket space V we introduce a dual space, called bra space, indicated with an asterisk as V^*, whose elements are in bi-univocal correspondence with those of V. In other words, we associate at each ket

$$|a\rangle \in V$$

the bra

$$\langle a| \in V^*,$$

uniquely

$$|a\rangle \longleftrightarrow \langle a|.$$

In the bra space the operations of sum and product for a scalar are similar to those in the ket space.

24.1 Sum of bras

The sum of two bras of the dual vectorial space V^* is still a bra belonging to the same space,

$$\forall \langle a|, \langle b| \in V^* \implies \langle a| + \langle b| \in V^*.$$

The sum of bras has the same properties of the sum of kets, i.e.:

Commutative property

Given two bras $\langle a|$ e $\langle b|$ we have the commutative property

$$\langle a| + \langle b| = \langle b| + \langle a|.$$

Associative property

Given a bra $\langle a|$ and two complex numbers $c_1, c_2 \in \mathbb{C}$ we have the associative property

$$\left(\langle a| + \langle b|\right) + \langle c| = \langle a| + \left(\langle b| + \langle c|\right).$$

24.2 Product for a scalar

The product of bra of a dual vectorial space V^* with a complex number is still a bra belonging to the same space

$$\forall \langle a| \in V^*, c \in \mathbb{C} \implies c \cdot \langle a| = \langle a| \cdot c \in V^*.$$

Usually we use the simplest notation $c\langle a|$ to indicate the product $c \cdot \langle a|$, analogously to the case of kets. In particular, the product of a ket for the number zero provides the null ket, in fact we can introduce the null element for the vectorial space (the null ket, indicated with $\langle 0|$), such that

$$\forall \, \langle a| \in V^* \implies 0\langle a| = \langle 0| \in V^*,$$
$$\forall \, \langle a| \in V^* \implies \langle a| + \langle 0| = \langle a|.$$

The ket product has the following properties:

Associative property

Given a bra $\langle a|$ and two complex numbers $c_1, c_2 \in \mathbb{C}$ we have the associative property

$$c_1 \cdot c_2\langle a| = c_1 \cdot (c_2\langle a|) = (c_1 c_2)\langle a|.$$

Distributive property with respect to scalars

Given a bra $\langle a|$ and two complex numbers $c_1, c_2 \in \mathbb{C}$ we have the distributive property with respect to scalars

$$(c_1 + c_2)\langle a| = c_1\langle a| + c_2\langle a|.$$

Distributive property with respect to kets

Given two bras $\langle a|$ and $\langle b|$ and a complex number $c \in \mathbb{C}$ we have the distributive property with respect to kets

$$c\Big(\langle a| + \langle b|\Big) = c\langle a| + c\langle b|\,.$$

24.3   Opposite of a bra

The opposite of bra $\langle a|$ is the bra $\langle b|$ such that

$$\langle a| + \langle b| = \langle 0|\,.$$

In particular, each bra of a dual vectorial space V^* admits its opposite ket which satisfies the previous relationship, i.e.

$$\forall\,\langle a| \in V^*,\ \exists\,\langle b| \in V^* : \langle a| + \langle b| = \langle 0|\,.$$

We indicate the opposite of bra $\langle a|$ with

$$-\langle a|\,,$$

from which

$$\langle a| - \langle a| = \langle 0|\,.$$

24.4 Difference of bras

The difference of two kets $\langle a|$ and $\langle b|$ of the dual vectorial space V^* is still a ket belonging to the same space

$$\forall \langle a|, \langle b| \in V^* \implies \langle a| - \langle b| \in V^*,$$

where the difference is defined as

$$\langle a| - \langle b| = \langle a| + (-\langle b|) = \langle a| + (-1)\langle b|,$$

obtained, as for the kets, by adding the opposite of the bra $|b\rangle$ to the bra $|a\rangle$. Moreover the commutative property does not generally apply to the difference

$$\langle a| - \langle b| \neq \langle b| - \langle a|.$$

24.5 Linear independence

The definition of linear independence for bras is analogous to that for kets, in particular, given n bras belonging to a dual vectorial space V^*, indicated with

$$\langle a_1|, \langle a_2|, \cdots, \langle a_n|,$$

we say that they are linearly independent if the equation

$$\sum_{k=1}^{n} c_k \langle a_k| = 0$$

has the only trivial solution

$$c_k = 0 \, , \forall k \, .$$

Otherwise it will be said that the n bras are linearly dependent.

24.6 Correspondence properties

We introduced, together with the kets of a vectorial space V, the bras of a dual space V^*, saying that they are related by a one-to-one correspondence. Given two kets $|a\rangle$ e $|b\rangle$ we have the following dual correspondences in sum or product for scalars

$$|a\rangle + |b\rangle \longleftrightarrow \langle a| + \langle b| \, ,$$
$$c|a\rangle \longleftrightarrow c^*\langle a| \, ,$$

where $c \in \mathbb{C}$ and c^* is the conjugate complex of c. These two properties are summarized in the generic dual correspondence

$$c_1 \cdot |a\rangle + c_2 \cdot |b\rangle \longleftrightarrow c_1^* \cdot \langle a| + c_2^* \cdot \langle b| \, , \qquad (24.6.1)$$

with $c_1, c_2 \in \mathbb{C}$.

Chapter 25

Internal product

The internal product between the two kets $|a\rangle$ and $|b\rangle$, also called scalar product, is a complex number and it is written as

$$\big(\langle a|\big) \cdot \big(|b\rangle\big) = \langle a|b\rangle \in \mathbb{C}\,.$$

We observe that the internal product is between the bra $\langle a|$, related to the ket $|a\rangle$, and the ket $|b\rangle$.

25.1 Fundamental properties

The first fundamental property for the internal product is the following: given two kets $|a\rangle$ e $|b\rangle$ we have

$$\langle a|b\rangle = \langle b|a\rangle^*\,. \tag{25.1.1}$$

In the case that

$$\langle b| = \langle a|,$$

it follows

$$\langle a|a \rangle = \langle a|a \rangle^* \implies \langle a|a \rangle \in \mathbb{R},$$

so the internal product of a ket with itself is always a real number. Starting from here we can exhibit the second property

$$\langle a|a \rangle \geq 0, \tag{25.1.2}$$

with the equality that holds if and only if $|a\rangle$ is the null ket, i.e.

$$\langle a|a \rangle = 0 \iff |a\rangle = 0.$$

Therefore

$$\langle a|a \rangle$$

is a non negative real number.

25.2   Linearity and antilinearity

Consider the kets $|a\rangle$ and the ket $|b\rangle$ of the vectorial space V, with

$$|b\rangle = \alpha|c\rangle + \beta|d\rangle \,, \quad |c\rangle, |d\rangle \in V \,,$$

and $\alpha, \beta \in \mathbb{C}$. Calculating the internal product we obtain

$$\langle a|b\rangle = \alpha\langle a|c\rangle + \beta\langle a|d\rangle \,,$$

i.e. it is linear with respect to the second factor. On the other hand, by calculating

$$\langle b| = \alpha^*\langle c| + \beta^*\langle d| \,,$$

from which

$$\langle b|a\rangle = \alpha^*\langle c|a\rangle + \beta^*\langle d|a\rangle \,,$$

i.e. the internal product is antilinear with respect to the first factor.

25.3   Norm

We define the norm of a ket $|a\rangle$ the real quantity

$$\sqrt{\langle a|a\rangle} \,.$$

Note that the root argument is non negative thanks to equation (25.1.2), i.e.

$$\langle a|a \rangle \geq 0 \,.$$

Using the fundamental postulate, it can be requested that the kets used to represent a physical state are normalized to 1, that is, have a norm equal to 1. Given a non zero ket $|a\rangle$, we can write the ket normalized to 1 given by

$$\frac{1}{\sqrt{\langle a|a \rangle}} \cdot |a\rangle \,.$$

The norm of ket $|a\rangle$ is often indicated with

$$\||a\rangle\| = \sqrt{\langle a|a \rangle} \,.$$

The norm in a metric space defines also the distance between two vectors. In fact, given two vectors $|a\rangle$ e $|b\rangle$, their distance is defined as

$$d\big(|a\rangle, |b\rangle\big) = \||a\rangle - |b\rangle\| \,.$$

25.4 Orthogonality

We can now define the concept of orthogonality between two state vectors. Two kets $|a\rangle$ and $|b\rangle$ are said orthogonal if

$$\langle a|b \rangle = 0 \,.$$

On other words two kets are orthogonal if the internal product between them is null, moreover, thanks to equation (25.1.1), i.e.

$$\langle a|b \rangle = \langle b|a \rangle^* \,,$$

is equivalent to saying that

$$\langle b|a \rangle = 0 \,.$$

The only ket that is orthogonal to itself is the null ket

$$\langle a|a \rangle = 0 \iff |a\rangle = 0 \,.$$

Moreover

$$\forall |b\rangle \in V, \langle a|b \rangle = 0 \iff |a\rangle = 0 \,,$$

i.e. the null ket is the only ket orthogonal to each other ket of the same vectorial space. Thanks to the definition of linear independence between kets we can say that a set of orthogonal kets that does not contain the null ket is made by linearly independent kets.

25.5 Schwarz inequality

The Schwarz inequality states that the modulus of the scalar product between two vectors is less than or equal to the product of the norms of the two vectors and is written

$$|\langle a|b\rangle| \leq \||a\rangle\| \cdot \||b\rangle\|,$$

or also

$$|\langle a|b\rangle| \leq \sqrt{\langle a|a\rangle}\sqrt{\langle b|b\rangle}.$$

To prove it, consider the vector

$$|c\rangle = |a\rangle + k\langle b|a\rangle|b\rangle, \quad k \in \mathbb{R}.$$

The bra corresponding to the ket c is $|c\rangle$, using equation (24.6.1), is

$$\langle c| = \langle a| + k\langle b|a\rangle^*\langle b|.$$

We compute the scalar product of the vector $|c\rangle$ with itself

$$\begin{aligned}
\langle c|c\rangle &= \Big(\langle a| + k\langle b|a\rangle^*\langle b|\Big)\Big(|a\rangle + k\langle b|a\rangle|b\rangle\Big) \\
&= \langle a|a\rangle + k\Big(\langle b|a\rangle^*\langle b|a\rangle + \langle b|a\rangle\langle a|b\rangle\Big) \\
&\quad + k^2\langle b|a\rangle^*\langle b|a\rangle\langle b|b\rangle.
\end{aligned}$$

Remembering that

$$\langle a|b\rangle = \langle b|a\rangle^*\,,$$

we obtain

$$
\begin{aligned}
\langle c|c\rangle &= \langle a|a\rangle + k\Big(|\langle b|a\rangle|^2 + |\langle b|a\rangle|^2\Big)\\
&+ k^2|\langle b|a\rangle|^2\langle b|b\rangle = \langle a|a\rangle + 2k|\langle b|a\rangle|^2\\
&+ k^2|\langle b|a\rangle|^2\langle b|b\rangle\,.
\end{aligned}
$$

This quantity is non negative, in fact from equation (25.1.2) we have

$$\langle c|c\rangle \geq 0\,,$$

therefore

$$\langle a|a\rangle + 2k|\langle b|a\rangle|^2 + k^2|\langle b|a\rangle|^2\langle b|b\rangle \geq 0\,,$$

for each value of the real parameter k. We can calculate the discriminant of the associated equation in k

$$\langle a|a\rangle + 2k|\langle b|a\rangle|^2 + k^2|\langle b|a\rangle|^2\langle b|b\rangle = 0\,,$$

from which

$$
\begin{aligned}
\Delta &= \Big(2|\langle b|a\rangle|^2\Big)^2 - 4|\langle b|a\rangle|^2\langle b|b\rangle\langle a|a\rangle\\
&= 4|\langle b|a\rangle|^2\Big(|\langle b|a\rangle|^2 - \langle b|b\rangle\langle a|a\rangle\Big)
\end{aligned}
$$

and put

$$\Delta \leq 0 \,,$$

so that the inequality is always satisfied, in fact in the cartesian plane of abscissa k, the first member represents a parabola with concavity upwards due to the coefficient of the second grad term which is always positive. Therefore we can write

$$4|\langle b|a\rangle|^2 \Big(|\langle b|a\rangle|^2 - \langle b|b\rangle\langle a|a\rangle \Big) \leq 0 \,,$$
$$|\langle b|a\rangle|^2 - \langle b|b\rangle\langle a|a\rangle \leq 0 \,,$$
$$|\langle b|a\rangle|^2 \leq \langle b|b\rangle\langle a|a\rangle \,,$$

hence the Schwarz inequality.

Chapter 26

Operators

Observables are physical quantities that can be measured. Mathematically we associate an operator to each observable. An X operator is indicated by the symbol $\hat{X}$. Measuring the physical quantity x for the physical state a means to apply, from the left, the operator associated with it, that is $\hat{X}$ on the state ket $|a\rangle \in V$, such that

$$\hat{X}|a\rangle \in V .$$

26.1 Action on ket and bra

An operator always acts to the left of a ket or to the right of a bra,

$$|a\rangle \in V \implies \hat{X}|a\rangle \in V$$

or

$$\langle a| \in V^* \implies \langle a|\hat{X} \in V^*$$

and each operator is uniquely defined by knowing its action on each vector. Moreover, two operators $\hat{X}$ e $\hat{Y}$ are equal if their action on an arbitrary ket is the same, i.e.

$$\hat{X} = \hat{Y} \iff \hat{X}|a\rangle = \hat{Y}|a\rangle, \quad \forall |a\rangle. \qquad (26.1.1)$$

26.2 Null operator

We can define the null operator as that operator $\hat{X}$ from which

$$\hat{X}|a\rangle = 0, \quad \forall |a\rangle,$$

i.e. its action on each ket gives the null ket.

26.3 Identity operator

The identity operator is indicated by $\hat{I}$ and satisfies the following relation

$$\hat{I}|a\rangle = |a\rangle, \quad \forall |a\rangle,$$

i.e. its action on each ket gives the same ket.

26.4 Eigenket and eigenvalues

The concepts of eigenket and eigenvalues are very important in physics.

Given an operator $\hat{X}$, the kets, indicated with

$$|\alpha^{(k)}\rangle, \quad k \in \mathbb{N},$$

such that

$$\hat{X}|\alpha^{(k)}\rangle = \alpha^{(k)}|\alpha^{(k)}\rangle, \quad k \in \mathbb{N},$$

with

$$\alpha^{(k)} \in \mathbb{C},$$

are called eigenkets of the operator $\hat{X}$ and the numbers $\alpha^{(k)}$ are said eigenvalues. The set

$$\{\alpha^{(k)}\}$$

is called set of eigenvalues for the operator $\hat{X}$. It is assumed that the autokets of a given observable form a basis in the N-dimensional space where they belong. So that any ket $|a\rangle$ of this space can be written as a linear combination of the basis kets, eigenkets of a certain operator

$\hat{X}$, in this way

$$|a\rangle = \sum_k c_k |\alpha^{(k)}\rangle ,$$

where

$$c_k \in \mathbb{C}$$

are appropriate coefficients. If an eigenvalue of a certain operator admits multiple eigenkets then that eigenvalue is said to be degenerate.

26.5 Sum of operators

The sum between operators satisfies the following properties:

Commutative property

Given two operators $\hat{X}$ and $\hat{Y}$ we have the commutative property

$$\hat{X} + \hat{Y} = \hat{Y} + \hat{X} .$$

Associative property

Given three operators $\hat{X}$, $\hat{Y}$ e $\hat{Z}$ we have the associative property

$$\hat{X} + (\hat{Y} + \hat{Z}) = (\hat{X} + \hat{Y}) + \hat{Z} \,.$$

26.6 Linearity and antilinearity

An operator X is said to be linear if it satisfies the property

$$\hat{X}\left(c_1|a\rangle + c_2|b\rangle\right) = c_1\hat{X}|a\rangle + c_2\hat{X}|b\rangle \,,$$

for each ket $|a\rangle$ and $|b\rangle$ and for each $c_1, c_2 \in \mathbb{C}$. Many operators in quantum mechanics satisfy this property. Similarly, an operator $\hat{X}$ is said to be antilinear if it satisfies the property

$$\hat{X}(c_1|a\rangle + c_2|b\rangle) = c_1^*\hat{X}|a\rangle + c_2^*\hat{X}|b\rangle \,,$$

for each ket $|a\rangle$ and $|b\rangle$ and for each $c_1, c_2 \in \mathbb{C}$. For example, in particle physics, the time reversal operator is antilinear.

26.7 Adjoint operator

Given an operator $\hat{X}$ and a ket $|a\rangle$ we define the adjoint operator to $\hat{X}$ (also called conjugated hermitian), indicated with

$$\hat{X}^\dagger,$$

the operator for which the bra

$$\langle a|\hat{X}^\dagger$$

is the dual of the ket

$$\hat{X}|a\rangle,$$

i.e.

$$\hat{X}|a\rangle \longleftrightarrow \langle a|\hat{X}^\dagger. \qquad (26.7.1)$$

26.8 Product of operators

The product of two operators is still an operator. Given two operators $\hat{X}$ and $\hat{Y}$ we have the associative property

$$\hat{X}(\hat{Y}\hat{Z}) = (\hat{X}\hat{Y})\hat{Z},$$

but the commutative one does not generally apply, in fact

$$\hat{X}\hat{Y} \neq \hat{Y}\hat{X}.$$

Moreover we have

$$\hat{X}(\hat{Y}|a\rangle) = (\hat{X}\hat{Y})|a\rangle = \hat{X}\hat{Y}|a\rangle$$
$$(\langle a|\hat{X})\hat{Y} = \langle a|(\hat{X}\hat{Y}) = \langle a|\hat{X}\hat{Y} .$$

26.9 Commutator and anticommutator

Since the product between operators is generally non--commutative we can define the commutator of two operators $\hat{X}$ and $\hat{Y}$ as

$$[\hat{X},\hat{Y}] = \hat{X}\hat{Y} - \hat{Y}\hat{X} .$$

If the commutator is vanishing

$$[\hat{X},\hat{Y}] = 0 ,$$

then the two operators are said to commutate.

We also define the anticommutator of the two operators $\hat{X}$ and $\hat{Y}$ as

$$\{\hat{X},\hat{Y}\} = \hat{X}\hat{Y} + \hat{Y}\hat{X} .$$

Analogously, if the anticommutator is vanishing

$$\{\hat{X},\hat{Y}\} = 0 ,$$

then the two operators are said to anticommute. Commutators satisfy the following properties

$$[\hat{X}, \hat{Y}] = -[\hat{Y}, \hat{X}],$$

$$[\hat{X}, \hat{Y}\hat{Z}] = \hat{Y}[\hat{X}, \hat{Z}] + [\hat{X}, \hat{Y}]\hat{Z},$$

$$[\hat{X}\hat{Y}, \hat{Z}] = \hat{X}[\hat{Y}, \hat{Z}] + [\hat{X}, \hat{Z}]\hat{Y},$$

$$\left[\hat{X}, [\hat{Y}, \hat{Z}]\right] + \left[\hat{Y}, [\hat{Z}, \hat{X}]\right] + \left[\hat{Z}, [\hat{X}, \hat{Y}]\right] = 0,$$

where the last expression is known also as the Jacobi identity.

26.10 Functions of operators

Thanks to the product definition, we can define the n-th power of an operator $\hat{X}$ in the following way

$$\hat{X}^n = \hat{X} \cdot \hat{X} \cdots \hat{X},$$

n times, with $n \geq 1$. Moreover

$$\hat{X}^0 = \hat{I},$$

where $\hat{I}$ is the identity operator. We can define the function of an operator using its Taylor series. For example we can write

$$e^{\hat{X}} = \sum_{k=0}^{\infty} \frac{\hat{X}^k}{k!}.$$

26.11 Inverse operator

Given an operator $\hat{X}$ we define its inverse (if it exists), denoted with

$$\hat{X}^{-1},$$

as the operator such that

$$\hat{X}\hat{X}^{-1} = \hat{X}^{-1}\hat{X} = \hat{I}.$$

We note that the identity operator is the reverse operator of himself

$$\hat{I}^{-1} = \hat{I}.$$

26.12 Adjoint of the product

Given two operators $\hat{X}$ and $\hat{Y}$ and a generic ket $|a\rangle$, consider the ket obtained from

$$\hat{X}\hat{Y}|a\rangle$$

and calculate the dual bra. Meanwhile, let's define the ket

$$|\gamma\rangle = \hat{Y}|a\rangle,$$

therefore

$$\hat{X}\hat{Y}|a\rangle = \hat{X}\big(\hat{Y}|a\rangle\big) = \hat{X}|\gamma\rangle\,.$$

We can use the equation (26.7.1), i.e.

$$\hat{X}|a\rangle \longleftrightarrow \langle a|\hat{X}^{\dagger}$$

and write

$$\hat{X}\hat{Y}|a\rangle \longleftrightarrow \langle\gamma|\hat{X}^{\dagger}\,. \qquad (26.12.1)$$

We calculate the bra $\langle\gamma|$ related to the ket $|\gamma\rangle$ using the equation (26.7.1), obtaining

$$|\gamma\rangle = \hat{Y}|a\rangle \longleftrightarrow \langle\gamma| = \langle a|\hat{Y}^{\dagger}\,,$$

and the equation (26.12.1) become

$$\hat{X}\hat{Y}|a\rangle \longleftrightarrow \langle\gamma|\hat{X}^{\dagger} = (\langle a|\hat{Y}^{\dagger})\hat{X}^{\dagger}\,,$$

or

$$\hat{X}\hat{Y}|a\rangle \longleftrightarrow \langle a|\hat{Y}^{\dagger}\hat{X}^{\dagger}\,.$$

Furthermore, we can write, using equation (26.7.1)

$$\hat{X}\hat{Y}|a\rangle = (\hat{X}\hat{Y})|a\rangle \longleftrightarrow \langle a|(\hat{X}\hat{Y})^{\dagger}$$

having considered

$$(\hat{X}\hat{Y})$$

as a single operator. Comparing with previous equation we see that the adjoint of the product of two operators is equal to the product of their adjoint in reverse order, i.e.

$$(\hat{X}\hat{Y})^\dagger = \hat{Y}^\dagger \hat{X}^\dagger .$$

Analogously for a multitude of operators

$$(\hat{A}\hat{B}\cdots\hat{Z})^\dagger = \hat{Z}^\dagger \cdots \hat{B}^\dagger \hat{A}^\dagger .$$

26.13 Hermitian operators

An operator $\hat{X}$ is said to be a hermitian (or self-adjoint) if it is equal to its adjoint, i.e.

$$\hat{X} = \hat{X}^\dagger .$$

The hermitian operators have important properties and they are usually used in quantum mechanics. First of all, the eigenvalues of a hermitian operator are real and the corresponding eigenkets of distinct eigenvalues are orthogonal. Indeed given a hermitian operator $\hat{X}$, if

$$|\alpha^{(k)}\rangle$$

are its eigenvalues and

$$|\alpha^{(k)}\rangle$$

its eigenkets we can write

$$\hat{X}|\alpha^{(k)}\rangle = \alpha^{(k)}|\alpha^{(k)}\rangle$$

and, multiplying[1] on the left by the bra

$$\langle\alpha^{(j)}|,$$

we obtain

$$\langle\alpha^{(j)}|\hat{X}|\alpha^{(k)}\rangle = \alpha^{(k)}\langle\alpha^{(j)}|\alpha^{(k)}\rangle. \qquad (26.13.1)$$

Using the equation (26.7.1), i.e.

$$\hat{X}|a\rangle \longleftrightarrow \langle a|\hat{X}^\dagger,$$

we can also write

$$\langle\alpha^{(j)}|\hat{X}^\dagger = \alpha^{(j)*}\langle\alpha^{(j)}|,$$

or also, being $\hat{X}$ hermitian by hypothesis,

$$\langle\alpha^{(j)}|\hat{X} = \alpha^{(j)*}\langle\alpha^{(j)}|,$$

by multiplying the latter expression by the ket

$$|\alpha^{(k)}\rangle$$

---

[1] see also the equation (26.15.1).

from the right, we get

$$\langle \alpha^{(j)} | \hat{X} | \alpha^{(k)} \rangle = \alpha^{(j)*} \langle \alpha^{(j)} | \alpha^{(k)} \rangle \,.$$

Comparing the latter with the equation (26.13.1), we have

$$\alpha^{(k)} \langle \alpha^{(j)} | \alpha^{(k)} \rangle = \alpha^{(j)*} \langle \alpha^{(j)} | \alpha^{(k)} \rangle \,,$$

or

$$(\alpha^{(k)} - \alpha^{(j)*}) \langle \alpha^{(j)} | \alpha^{(k)} \rangle = 0 \,.$$

Choosing $j = k$ we obtain

$$(\alpha^{(k)} - \alpha^{(k)*}) \langle \alpha^{(k)} | \alpha^{(k)} \rangle = 0$$

and using

$$\langle a | a \rangle = 0 \iff |a\rangle = 0 \,,$$

the number

$$\langle \alpha^{(k)} | \alpha^{(k)} \rangle$$

is vanishing if and only if the ket

$$|\alpha^{(k)}\rangle$$

is the null one, but this is not the case, therefore necessarily

$$\alpha^{(k)} - \alpha^{(k)*} = 0 \,,$$

from which

$$\alpha^{(k)} \in \mathbb{R} \,, \quad \forall \, k \,.$$

On the other hand, if we take distinct eigenvalues, i.e.

$$(\alpha^{(k)} - \alpha^{(j)*}) \neq 0 \,,$$

then it immediately follows that

$$\langle \alpha^{(j)} | \alpha^{(k)} \rangle = 0$$

i.e., the autokets $\langle \alpha^{(j)} |$ and $| \alpha^{(k)} \rangle$ are orthogonal to each other, with $j \neq k$. In general these eigenkets are normalized to 1 (with an appropriate choice of the arbitrary multiplicative constant of a ket), so that they form a set of orthonormal kets, i.e. such that

$$\langle \alpha^{(k)} | \alpha^{(j)} \rangle = \delta_{kj} \,, \qquad (26.13.2)$$

with δ_{kj} is the Kronecker delta defined by

$$\delta_{kj} := \begin{cases} 0 & \text{if } k \neq j \\ 1 & \text{if } k = j \end{cases}.$$

If $\hat{X}$ and $\hat{Y}$ are two hermitian operators we have

$$(\hat{X}\hat{Y})^\dagger = \hat{Y}^\dagger \hat{X}^\dagger = \hat{Y}\hat{X},$$

which shows that the product of two hermitian operators is hermitian if and only if they commutate.

Finally an operator is said to be antihermitian if

$$\hat{X}^\dagger = -\hat{X}.$$

26.14   Unitary operators

An operator U is said to be unitary if

$$\hat{U}\hat{U}^\dagger = \hat{U}^\dagger \hat{U} = \hat{I}.$$

Let's see an important property of the unit operators. We write, given a generic ket $|a\rangle$ and a unitary operator U, the ket transformed by U

$$|a'\rangle = \hat{U}|a\rangle$$

and the dual bra

$$\langle a' | = \langle a | \hat{U}^\dagger .$$

By multiplying[2] the latter bra to the left with the previous ket we obtain

$$\langle a' | a \rangle = \langle a | \hat{U}^\dagger \hat{U} | a \rangle$$

and, from the definition of unitary operator, it follows that

$$\langle a' | a' \rangle = \langle a | \hat{U}^\dagger \hat{U} | a \rangle = \langle a | \hat{I} | a \rangle .$$

Since the action of the identity operator on a ket leaves it unchanged, from the identity

$$\hat{I} | a \rangle = | a \rangle , \quad \forall \, | a \rangle$$

we come to

$$\langle a' | a' \rangle = \langle a | a \rangle .$$

A unitary operator, when transforming kets, leaves the norm unchanged, being

$$\langle a | a \rangle \geq 0 .$$

---

[2] see also the equation (26.15.1).

26.15   External product

The external product between a ket $\langle b|$ and a bra $|a\rangle$, is indicated with

$$\big(|b\rangle\big) \cdot \big(\langle a|\big) = |b\rangle\langle a|$$

and represents an operator with the following property

$$\big(|b\rangle\langle a|\big)^{\dagger} = |a\rangle\langle b| \, .$$

26.15.1   Associativity

We postulate that the multiplication operations between ket and bra are associative. In the particular case of an external product this means that

$$\big(|b\rangle\langle a|\big) \cdot |\beta\rangle = |b\rangle\big(\langle a|\beta\rangle\big) = \langle a|\beta\rangle|b\rangle \, ,$$

with

$$\langle a|\beta\rangle \in \mathbb{C} \, .$$

So the action of the external product on the ket

$$|b\rangle\langle a|$$

produces a new ket. Therefore the external product acts as an operator. Moreover, given an operator $\hat{X}$ and two kets $|a\rangle$ and $|b\rangle$, the associativity implies that

$$\left(\langle b|\right) \cdot \left(\hat{X}|a\rangle\right) = \left(\langle b|\hat{X}\right) \cdot |a\rangle = \langle b|\hat{X}|a\rangle. \qquad (26.15.1)$$

In particular we can write, using the equation (25.1.1)

$$\langle a|b\rangle = \langle b|a\rangle^*,$$

from which

$$\langle b|\hat{X}|a\rangle = \langle b|(\hat{X}|a\rangle) = \langle \gamma|b\rangle^*,$$

where $\langle \gamma|$ is the dual bra of the ket

$$\hat{X}|a\rangle,$$

i.e.

$$\langle \gamma| \longleftrightarrow \hat{X}|a\rangle.$$

From the equation (26.7.1)

$$\hat{X}|a\rangle \longleftrightarrow \langle a|\hat{X}^\dagger$$

it follows that

$$\langle \gamma| = \langle a|\hat{X}^\dagger$$

and the previous equation becomes

$$\langle b|\hat{X}|a\rangle = \langle a|\hat{X}^\dagger|b\rangle^*.$$

26.15.2 Projectors

A projector $\hat{P}$ is a hermitian and idempotent operator, i.e., it satisfies the following conditions

$$\hat{P}^\dagger = \hat{P},$$
$$\hat{P}^2 = \hat{P}.$$

The identity operator is an example of projector and it is the only one to admit reverse. An example of a projector is the following operator

$$\hat{P} = |a\rangle\langle a|,$$

with $|a\rangle$ unitary vector,

$$\langle a|a\rangle = 1.$$

In fact it is hermitian

$$\hat{P}^\dagger = (|a\rangle\langle a|)^\dagger = |a\rangle\langle a| = \hat{P}$$

and idempotent

$$\hat{P}^2 = (|a\rangle\langle a|)(|a\rangle\langle a|) = |a\rangle\langle a|a\rangle\langle a| = |a\rangle\langle a| = \hat{P}.$$

The action of this projector on a generic vector $|b\rangle$ is

$$\hat{P}|b\rangle = (|a\rangle\langle a|)|b\rangle = \langle a|b\rangle|a\rangle,$$

where the quantity

$$\langle a|b \rangle$$

can be seen as the "projection" of the vector b in the direction of the unitary vector a.

26.16 Eigenkets basis

The eigenkets of a hermitian operator $\hat{A}$ form a complete set. With unitary normalization this becomes a complete orthonormal set. Therefore any ket $|a\rangle$ can be written as a single linear combination, with complex coefficients, of the eigenkets of $\hat{A}$. We can write

$$|a\rangle = \sum_j c_j |\alpha^{(j)}\rangle, \qquad (26.16.1)$$

with $c_j \in \mathbb{C}$. By multiplying on the left with the bra

$$\langle \alpha^{(k)}|$$

of a generic eigenket

$$|\alpha^{(k)}\rangle$$

of $\hat{A}$ and using the equation (26.13.2)

$$\langle \alpha^{(k)}|\alpha^{(j)}\rangle = \delta_{k,j} ,$$

we have

$$\langle \alpha^{(k)} | a \rangle = \sum_j c_j \langle \alpha^{(k)} | \alpha^{(j)} \rangle = \sum_j c_j \delta_{k,j} = c_k \, ,$$

or

$$c_k = \langle \alpha^{(k)} | a \rangle \, .$$

The coefficient c_j in the equation (26.16.1) is therefore given by

$$\langle \alpha^{(j)} | a \rangle$$

and we can write

$$|a\rangle = \sum_k \langle \alpha^{(k)} | a \rangle \cdot |\alpha^{(k)}\rangle = \sum_k |\alpha^{(k)}\rangle \cdot \langle \alpha^{(k)} | a \rangle \, .$$

From associativity this expression can be also written as

$$
\begin{aligned}
|a\rangle &= \sum_k \left(|\alpha^{(k)}\rangle \langle \alpha^{(k)}| \right) |a\rangle \\
&= \left(\sum_k |\alpha^{(k)}\rangle \langle \alpha^{(k)}| \right) |a\rangle \, ,
\end{aligned}
$$

valid for any arbitrary ket $|a\rangle$. So thanks to the definition in the equation (26.1.1),

$$\hat{X} = \hat{Y} \iff \hat{X}|a\rangle = \hat{Y}|a\rangle \, , \quad \forall |a\rangle \, ,$$

we can write the operator equality

$$\sum_k |\alpha^{(k)}\rangle\langle\alpha^{(k)}| = \hat{I},$$

said completeness relation. As a first use of this identity let's start by

$$\langle a|a\rangle = \langle a|\hat{I}|a\rangle,$$

from which

$$
\begin{aligned}
\langle a|a\rangle &= \langle a| \left(\sum_k |\alpha^{(k)}\rangle\langle\alpha^{(k)}| \right) |a\rangle \\
&= \sum_k \langle a|\alpha^{(k)}\rangle\langle\alpha^{(k)}|a\rangle.
\end{aligned}
$$

Using the equation (25.1.1)

$$\langle a|b\rangle = \langle b|a\rangle^*,$$

we have

$$
\begin{aligned}
\langle a|a\rangle &= \sum_k \langle\alpha^{(k)}|a\rangle\langle\alpha^{(k)}|a\rangle^* \\
&= \sum_k |\langle\alpha^{(k)}|a\rangle|^2 = \sum_k |c_k|^2,
\end{aligned}
$$

where we have used

$$c_k = \langle\alpha^{(k)}|a\rangle.$$

In particular if the ket $|a\rangle$ is normalized to 1, then the coefficients satisfy the relation

$$\sum_k |c_k|^2 = 1.$$

Chapter 27

Representation of vectors and operators

27.1 Representation of a ket

Given a vectorial space V of dimension N, consider the basis composed by the N linear independent vectors

$$|e_i\rangle, \quad i = 1, 2, \cdots, N.$$

A generic vector $|a\rangle$ can be written in this space as a linear combination of the elements of the V basis as

$$|a\rangle = \sum_{i=1}^{N} a^i |e_i\rangle,$$

where the coefficients

$$a^i, \quad i = 1, 2, \cdots, N$$

are the components of the vector $|a\rangle$ in the above basis. The vector $|a\rangle$ can therefore be represented by a one-column matrix with N rows (column vector) as

$$a = \begin{pmatrix} a^1 \\ a^2 \\ \vdots \\ a^N \end{pmatrix}.$$

27.2 Representation of an operator

Similarly, also an operator can be represented by a matrix on a certain basis. Consider a generic operator $\hat{P}$ acting on the vector of the V space basis

$$\hat{P}|e_i\rangle.$$

Assuming that the resulting vector belongs to the same vectorial space, i.e. that $\hat{P}$ transforms vectors of V into vectors of V, the resulting vector will be expressed as a linear combination of the basis vectors as

$$\hat{P}|e_i\rangle = \sum_{j=1}^{N} P_i^{\ j}|e_j\rangle.$$

The components have two indices (i and j) and can be viewed as elements of a square matrix of N rows and N

columns, as follows

$$P = \begin{pmatrix} P_1^1 & P_2^1 & \cdots & P_N^1 \\ P_2^1 & P_2^2 & \cdots & P_N^2 \\ \vdots & \vdots & \ddots & \vdots \\ P_N^1 & P_2^N & \cdots & P_N^N \end{pmatrix}.$$

This matrix can be seen as the representation of the operator $\hat{P}$ in the chosen basis.

27.2.1 Operations between matrices

Product of matrices

The algebra of matrices is very similar to that of vectors. In particular, the product between two matrices is defined as "row by column multiplication". Given a matrix A of size $N \times M$ and a matrix B of size $M \times L$, we consider the product matrix C (matrix of size $N \times L$)

$$C = A \cdot B \,,$$

whose elements are

$$C_j^i = \sum_{k=1}^{M} A_k^i B_j^k \,,$$

with

$$i = 1, 2, \cdots, N, \quad j = 1, 2, \cdots, L.$$

Given a vector $|a\rangle$ and an operator $\hat{P}$, represented, respectively, by a column vector and a matrix, choose an appropriate basis, we can use the algebra of matrices to express the vector $|b\rangle$ obtained by the application of $\hat{P}$ on $|a\rangle$, i.e.

$$|b\rangle = \hat{P}|a\rangle,$$

as

$$b = P \cdot a,$$

or

$$\begin{pmatrix} b^1 \\ b^2 \\ \vdots \\ b^N \end{pmatrix} = \begin{pmatrix} P_1^1 & P_2^1 & \cdots & P_N^1 \\ P_2^1 & P_2^2 & \cdots & P_N^2 \\ \vdots & \vdots & \ddots & \vdots \\ P_N^1 & P_2^N & \cdots & P_N^N \end{pmatrix} \begin{pmatrix} a^1 \\ a^2 \\ \vdots \\ a^N \end{pmatrix},$$

where, in a compact form,

$$b^i = \sum_{j=1}^{N} P_j^i a^j.$$

Determinant

For a 2×2 square matrix A

$$A = \begin{pmatrix} a_{11} & a_{12} \\ a_{21} & a_{22} \end{pmatrix},$$

the determinant is given by

$$\det(A) = a_{11}a_{22} - a_{12}a_{21}$$

that can be written also as

$$\det(A) = \begin{vmatrix} a_{11} & a_{12} \\ a_{21} & a_{22} \end{vmatrix}.$$

For a 3×3 matrix

$$A = \begin{pmatrix} a_{11} & a_{12} & a_{13} \\ a_{21} & a_{22} & a_{23} \\ a_{31} & a_{32} & a_{33} \end{pmatrix},$$

we have

$$\det(A) = a_{11} \begin{vmatrix} a_{22} & a_{23} \\ a_{32} & a_{33} \end{vmatrix} - a_{12} \begin{vmatrix} a_{21} & a_{23} \\ a_{31} & a_{33} \end{vmatrix}$$
$$+ a_{13} \begin{vmatrix} a_{21} & a_{22} \\ a_{31} & a_{32} \end{vmatrix},$$

where the determinants of the 2×2 sub-matrices shown in the second side of the equation are obtained, for each term, by eliminating from the original 3×3 matrix the row and column corresponding to the first element chosen in the product. For example, for the first term

$$a_{11} \begin{vmatrix} a_{22} & a_{23} \\ a_{32} & a_{33} \end{vmatrix},$$

having chosen the element a_{11} it is necessary to multiply it by the determinant of the 2×2 sub-matrix which is obtained by eliminating from matrix A the row and column corresponding to the element a_{11}, i.e.

$$\begin{vmatrix} a_{22} & a_{23} \\ a_{32} & a_{33} \end{vmatrix}.$$

For the second term

$$-a_{12} \begin{vmatrix} a_{21} & a_{23} \\ a_{31} & a_{33} \end{vmatrix},$$

we pass to the element a_{12} (for each switch it is necessary to introduce a minus sign) and we should multiply it by the determinant of the 2×2 sub-matrix that is obtained by eliminating from A the row and column corresponding to the element a_{12}, i.e.

$$\begin{vmatrix} a_{21} & a_{23} \\ a_{31} & a_{33} \end{vmatrix}$$

and so on, moving towards the rightmost element and each time changing sign. This procedure is valid for calculating the determinant of any square matrix NxN. In this way it is possible to write the determinant of a $N \times N$ matrix in function of the determinants of the N sub-matrices $(N-1) \times (N-1)$.

The determinant is an invariant, i.e. it does not depend on the basis chosen to represent the matrix.

Trace

The trace of a square matrix is defined as the sum of the elements placed on the main diagonal. Let's consider the $N \times N$ matrix

$$A = \begin{pmatrix} a_{11} & a_{12} & \cdots & a_{1N} \\ a_{21} & a_{22} & \cdots & a_{2N} \\ \vdots & \vdots & \ddots & \vdots \\ a_{N1} & a_{N2} & \cdots & a_{NN} \end{pmatrix},$$

the trace is defined by

$$\mathrm{Tr}(A) = \sum_{i=1}^{N} a_{ii}.$$

The trace of the product of multiple $N \times N$ matrices is invariant if the matrices are cyclically permutated, i.e. you can move the first matrix to the last position (and vice versa) as many times as you want without changing the result. For example, given three matrices A, B and C, we can write

$$\text{Tr}(ABC) = \text{Tr}(BCA) = \text{Tr}(CAB).$$

To prove it, remember that

$$ABC = A(BC) = \sum_{k=1}^{N} a_{ik}(BC)_{kj},$$

with

$$(BC)_{kj} = \sum_{n=1}^{N} b_{kn}c_{nj},$$

from which

$$ABC = \sum_{k=1}^{N}\sum_{n=1}^{N} a_{ik}b_{kn}c_{nj},$$

and

$$\text{Tr}(ABC) = \sum_{i=1}^{N}(ABC)_{ii}$$
$$= \sum_{i=1}^{N}\sum_{k=1}^{N}\sum_{n=1}^{N} a_{ik}b_{kn}c_{ni}.$$

We observe that the elements of the matrices can be changed in position in the last member, being elements and not matrices and satisfy the commutative property.

So we can also write, for example,

$$\text{Tr}(ABC) = \sum_{i=1}^{N}(ABC)_{ii}$$

$$= \sum_{i=1}^{N}\sum_{k=1}^{N}\sum_{n=1}^{N} c_{ni}a_{ik}b_{kn} \, ,$$

where we recognize the product of the three matrices in the CAB order, that is

$$\text{Tr}(ABC) = \sum_{i=1}^{N}\sum_{k=1}^{N}\sum_{n=1}^{N} c_{ni}a_{ik}b_{kn}$$

$$= \sum_{i=1}^{N}(CAB)_{ii} \, .$$

The matrix elements are in the correct order to be interpreted as a product of 3 matrices only if we move the last factor from the last to the first position or vice versa as many time as we want.

The trace is an invariant, i.e. it does not depend on the basis chosen to represent the matrix.

27.3 Representation of a bra

Given a vectorial space V of dimension N, consider the basis composed by the N linear independent vectors

$$|e_i\rangle , \quad i = 1, 2, \cdots, N \, .$$

We have seen previously that a generic ket $|a\rangle$ can be written as

$$|a\rangle = \sum_{i=1}^{N} a^i |e_i\rangle \,,$$

with coefficients

$$a^i \,, \quad i = 1, 2, \cdots, N \,.$$

For the bra we can write

$$\langle a| = \sum_{i=1}^{N} a^{i*} \langle e_i| \,.$$

Consider now another vector

$$|b\rangle = \sum_{i=1}^{N} b^i |e_i\rangle$$

and calculate the scalar product between $|a\rangle$ and $|b\rangle$

$$
\begin{aligned}
\langle a|b\rangle &= \left(\sum_{j=1}^{N} a^{j*} \langle e_j| \right) \left(\sum_{i=1}^{N} b^i |e_i\rangle \right) \\
&= \sum_{i=1}^{N} \sum_{j=1}^{N} a^{j*} b^i \langle e_j|e_i\rangle \,,
\end{aligned}
$$

that we can write also as

$$\langle a|b\rangle = \sum_{i=1}^{N} a_i b^i \,,$$

having defined

$$a_i = \sum_{j=1}^{N} a^{j*} \langle e_j | e_i \rangle \,,$$

which are the components of a row vector representing the bra vector a, i.e.

$$a^\dagger = \begin{pmatrix} a_1 & a_2 & \cdots & a_N \end{pmatrix} \,.$$

In this way the scalar product between two vectors $|a\rangle$ and $|b\rangle$ can be simply written as

$$\langle a | b \rangle = \sum_{i=1}^{N} a_i b^i = a^\dagger \cdot b$$

or

$$\langle a | b \rangle = \begin{pmatrix} a_1 & a_2 & \cdots & a_N \end{pmatrix} \begin{pmatrix} b^1 \\ b^2 \\ \vdots \\ b^N \end{pmatrix} \,.$$

Part IV

The Dirac equation

Introduction

This book is dedicated to the Dirac equation. The main
arguments are: Dirac equation, gamma matrices in Dirac
representation, properties of gamma matrices, covariance
of the Dirac equation, Dirac Lagrangian and derivation
of the Dirac equation from the equations of the Euler-Lagrange motion, Dirac equation in Hamiltonian form and
free solutions in the rest and in any reference frame.

Chapter 28

Notations

Generally we will use the Greek letters to indicate the indices that take values $0, 1, 2, 3$ and the Latin ones for the indices that assume values $1, 2, 3$. Moreover we will use the Minkowski's metric tensor with signature $(+, -, -, -)$, i.e.

$$\eta = \begin{pmatrix} 1 & 0 & 0 & 0 \\ 0 & -1 & 0 & 0 \\ 0 & 0 & -1 & 0 \\ 0 & 0 & 0 & -1 \end{pmatrix}$$

and the Einstein summation convention that implies summation over two repeated indexes, one contravariant and one covariant, as for example,

$$q^\mu p_\mu = \sum_{\mu=0}^{3} q^\mu p_\mu \,.$$

Finally we will use the natural units, where

$$\hbar = c = 1 \, .$$

Chapter 29

Dirac equation

29.1 The Dirac equation

The Dirac equation can be written, in natural units, in the form

$$(i\gamma^\mu \partial_\mu - m)\psi(x) = 0\,,$$

where i it is the imaginary unity, with

$$i^2 = -1$$

and where the γ^μ are four 4×4 constant matrices (called Dirac gamma matrices), one for each value that the Greek letter μ can take

$$\mu = 0, 1, 2, 3\,.$$

The wave function (spinor) is

$$\psi(x) = \begin{pmatrix} \psi^1(x) \\ \psi^2(x) \\ \psi^3(x) \\ \psi^4(x) \end{pmatrix},$$

where x is the position four-vector. Writing explicitly the sum implied by the Einstein convention, the Dirac equation can be written as

$$\left(i \sum_{\mu=0}^{3} \gamma^\mu \partial_\mu - m \right) \psi(x) = 0,$$

or also, by writing the product between matrices,

$$i \sum_{\mu=0}^{3} \sum_{\beta=0}^{3} (\gamma^\mu)_{\alpha\beta} \partial_\mu \psi^\beta(x) - m\psi^\alpha(x) = 0,$$

$$\forall \alpha = 0, 1, 2, 3.$$

Note that the term

$$(\gamma^\mu)_{\alpha\beta},$$

is the element of row α and column β of the μ-th Dirac gamma matrix.

29.2 Commutation relations for the γ matrices

Consider the Dirac equation in the form

$$(i\gamma^\mu \partial_\mu - m)\psi(x) = 0$$

and multiply it to the left by

$$i\gamma^\nu \partial_\nu + m \,,$$

we obtain

$$(i\gamma^\nu \partial_\nu + m)(i\gamma^\mu \partial_\mu - m)\psi(x) = 0 \,.$$

Calculating

$$(-\gamma^\nu \gamma^\mu \partial_\mu \partial_\nu + im\gamma^\mu \partial_\mu - im\gamma^\nu \partial_\nu - m^2)\psi(x) = 0 \,,$$

from which

$$(-\gamma^\nu \gamma^\mu \partial_\mu \partial_\nu - m^2)\psi(x) = 0 \,.$$

In fact we remember that

$$im\gamma^\mu \partial_\mu = im\gamma^\nu \partial_\nu \,,$$

due to the implied sum on the indices,

$$im \sum_{\mu=0}^{3} \gamma^\mu \partial_\mu = im \sum_{\nu=0}^{3} \gamma^\nu \partial_\nu \,.$$

Therefore we obtain the equation

$$(\gamma^\nu \gamma^\mu \partial_\mu \partial_\nu + m^2)\psi(x) = 0 \,,$$

which is identical to

$$\left(\frac{\gamma^\nu \gamma^\mu + \gamma^\mu \gamma^\nu + \gamma^\nu \gamma^\mu - \gamma^\mu \gamma^\nu}{2} \partial_\mu \partial_\nu + m^2 \right) \psi(x) = 0 \,.$$

Remember that for the product of Dirac gamma matrices, being matrices, the commutative property does not apply and therefore

$$\gamma^\nu \gamma^\mu \neq \gamma^\mu \gamma^\nu \,.$$

Using the commutator and anti-commutator operators between matrices, defined respectively by

$$[\gamma^\mu, \gamma^\nu] := \gamma^\mu \gamma^\nu - \gamma^\nu \gamma^\mu$$

and

$$\{\gamma^\mu, \gamma^\nu\} := \gamma^\mu \gamma^\nu + \gamma^\nu \gamma^\mu \,,$$

the equation becomes

$$\left(\frac{[\gamma^\mu, \gamma^\nu] + \{\gamma^\mu, \gamma^\nu\}}{2} \partial_\mu \partial_\nu + m^2 \right) \psi(x) = 0 \,.$$

Note now that the term

$$[\gamma^\mu, \gamma^\nu] \partial_\mu \partial_\nu = \sum_{\mu=0}^{3} \sum_{\nu=0}^{3} [\gamma^\mu, \gamma^\nu] \partial_\mu \partial_\nu = 0$$

is identically null, being a contraction between a symmetric term for the exchange of the indices and an antisymmetric term, in fact we can write

$$
\begin{aligned}
[\gamma^\mu, \gamma^\nu] \partial_\mu \partial_\nu &= \sum_{\mu=0}^{3} \sum_{\nu=0}^{3} (\gamma^\mu \gamma^\nu - \gamma^\nu \gamma^\mu) \partial_\mu \partial_\nu \\
&= \sum_{\mu=0}^{3} \sum_{\nu=0}^{3} \gamma^\mu \gamma^\nu \partial_\mu \partial_\nu \\
&\quad - \sum_{\mu=0}^{3} \sum_{\nu=0}^{3} \gamma^\nu \gamma^\mu \partial_\mu \partial_\nu \\
&= \sum_{\mu=0}^{3} \sum_{\nu=0}^{3} \gamma^\mu \gamma^\nu \partial_\mu \partial_\nu \\
&\quad - \sum_{\mu=0}^{3} \sum_{\nu=0}^{3} \gamma^\mu \gamma^\nu \partial_\nu \partial_\mu \,,
\end{aligned}
$$

where in the last passage we renamed the indices with a summation implied, from which finally

$$[\gamma^\mu, \gamma^\nu]\,\partial_\mu\partial_\nu \;=\; \sum_{\mu=0}^{3}\sum_{\nu=0}^{3}\gamma^\mu\gamma^\nu\partial_\mu\partial_\nu$$

$$-\;\sum_{\mu=0}^{3}\sum_{\nu=0}^{3}\gamma^\mu\gamma^\nu\partial_\mu\partial_\nu = 0\,.$$

In light of this result, the previous equation can be simply written as

$$\left(\frac{1}{2}\,\{\gamma^\mu, \gamma^\nu\}\,\partial_\mu\partial_\nu + m^2\right)\psi(x) = 0\,.$$

To relate this equation to the known Klein-Gordon equation

$$(\partial^\mu\partial_\mu + m^2)\psi = 0\,,$$

i.e.

$$(\eta^{\mu\nu}\partial_\mu\partial_\nu + m^2)\psi = 0\,,$$

it must be necessarily

$$\frac{1}{2}\,\{\gamma^\mu, \gamma^\nu\} = \eta^{\mu\nu}I\,,$$

or

$$\gamma^\mu\gamma^\nu + \gamma^\nu\gamma^\mu = 2\eta^{\mu\nu}I\,,$$

where I is the identity matrix 4×4 which we will omit from now on and we will understand from the context if we are manipulating matrices or numbers (for example if we add a matrix to a number we mean that number is multiplied by the identity matrix I). Note how this assumption was already made when we wrote the Dirac equation where a matrix operator was added to the opposite of the mass which is a scalar.

The previous relations (since the equation includes 16 equations, one for each choice of the free indices) characterize the Dirac gamma matrices and it is possible to derive various properties from them.

Meanwhile, since the metric tensor assumes null values when the two indices are different (the elements outside the diagonal are null), we immediately obtain that two distinct gamma matrices anti-commute, i.e.

$$\{\gamma^\mu, \gamma^\nu\} = 0, \quad \mu \neq \nu.$$

The anti-commutator of two identical matrices depends instead on the index. If we choose the temporal component (index 0) we have

$$\{\gamma^0, \gamma^0\} = 2(\gamma^0)^2 = 2\eta^{00} = 2,$$

from which

$$(\gamma^0)^2 = 1\,,$$

while for spatial indices (indices 1,2,3, indicated with Latin letters)

$$\{\gamma^j, \gamma^j\} = 2(\gamma^j)^2 = 2\eta^{jj} = -2\,,$$

i.e.

$$(\gamma^j)^2 = -1\,, \quad j = 1, 2, 3\,.$$

29.3   The continuity equation

We denote with the symbol $\psi^\dagger$ the conjugated Hermitian of the spinor ψ. Let's write the Dirac equation

$$(i\gamma^\mu \partial_\mu - m)\psi = 0\,,$$

or

$$i\gamma^\mu \partial_\mu \psi - m\psi = 0$$

and consider his conjugated Hermitian

$$-i(\partial_\mu \psi^\dagger)\gamma^{\mu\dagger} - m\psi^\dagger = 0\,.$$

It should be noted that the partial derivative acts on the spinor and that for the Hermitian conjugate of a generic product AB we have

$$(AB)^\dagger = B^\dagger A^\dagger$$

and remember that the conjugated Hermitian of a complex number coincides with its conjugate complex, therefore

$$i^\dagger = i^* = -i \,.$$

Multiply the initial Dirac equation by the quantity

$$\psi^\dagger \gamma^0 \,,$$

from left, obtaining

$$i\psi^\dagger \gamma^0 \gamma^\mu \partial_\mu \psi - m\psi^\dagger \gamma^0 \psi = 0$$

and the conjugated Hermitian equation of the Dirac equation for the quantity

$$\gamma^{0\dagger} \psi \,,$$

from the right, from which

$$-i(\partial_\mu \psi^\dagger)\gamma^{\mu\dagger}\gamma^{0\dagger}\psi - m\psi^\dagger \gamma^{0\dagger}\psi = 0 \,.$$

We subtract the latter equation from the previous one, obtaining

$$i\psi^\dagger\gamma^0\gamma^\mu\partial_\mu\psi - m\psi^\dagger\gamma^0\psi + i(\partial_\mu\psi^\dagger)\gamma^{\mu\dagger}\gamma^{0\dagger}\psi$$
$$+m\psi^\dagger\gamma^{0\dagger}\psi = 0\,,$$

i.e.

$$i\left(\psi^\dagger\gamma^0\gamma^\mu\partial_\mu\psi + (\partial_\mu\psi^\dagger)\gamma^{\mu\dagger}\gamma^{0\dagger}\psi\right)$$
$$+m(\psi^\dagger\gamma^{0\dagger}\psi - \psi^\dagger\gamma^0\psi) = 0\,.$$

We want to obtain a continuity equation, represented by a four-divergence contracted with a four-current equal to zero

$$\partial_\mu J^\mu = 0\,.$$

We must eliminate the second term containing the mass m, assuming that the matrix γ^0 is Hermitian, i.e.

$$\gamma^{0\dagger} = \gamma^0\,,$$

from which the previous equation becomes

$$\psi^\dagger\gamma^0\gamma^\mu\partial_\mu\psi + (\partial_\mu\psi^\dagger)\gamma^{\mu\dagger}\gamma^{0\dagger}\psi = 0\,.$$

Observe that we could write the first member as

$$\partial_\mu(\psi^\dagger\gamma^0\gamma^\mu\psi) = \psi^\dagger\gamma^0\gamma^\mu\partial_\mu\psi + (\partial_\mu\psi^\dagger)\gamma^0\gamma^\mu\psi\,,$$

if

$$\gamma^{\mu\dagger}\gamma^{0\dagger} = \gamma^0\gamma^\mu,$$

or if the product

$$\gamma^0\gamma^\mu$$

were Hermitian, in fact

$$(\gamma^0\gamma^\mu)^\dagger = \gamma^{\mu\dagger}\gamma^{0\dagger}.$$

Under this hypothesis we obtain the continuity equation

$$\partial_\mu(\psi^\dagger\gamma^0\gamma^\mu\psi) = 0$$

and we can define the four-current of probability

$$J^\mu = \psi^\dagger\gamma^0\gamma^\mu\psi = \overline{\psi}\gamma^\mu\psi,$$

where we have introduced the adjoint spinor defined as

$$\overline{\psi} = \psi^\dagger\gamma^0.$$

From the expression

$$\gamma^0\gamma^\mu = \gamma^{\mu\dagger}\gamma^{0\dagger},$$

remembering that

$$\gamma^{0\dagger} = \gamma^0 \,,$$

we have

$$\gamma^0 \gamma^\mu = \gamma^{\mu\dagger} \gamma^0$$

and multiplying to the left by γ^0 we get

$$\gamma^0 \gamma^0 \gamma^\mu = \gamma^0 \gamma^{\mu\dagger} \gamma^0 \,,$$
$$(\gamma^0)^2 \gamma^\mu = \gamma^0 \gamma^{\mu\dagger} \gamma^0 \,,$$
$$\gamma^\mu = \gamma^0 \gamma^{\mu\dagger} \gamma^0 \,,$$

where we have used

$$(\gamma^0)^2 = 1 \,,$$

from which also

$$\gamma^{\mu\dagger} = \gamma^0 \gamma^\mu \gamma^0 \,.$$

Finally, we observe that the temporal component of the four-current is

$$J^0 = \overline{\psi} \gamma^0 \psi = \psi^\dagger (\gamma^0)^2 \psi = \psi^\dagger \psi \,.$$

Chapter 30

Properties of the γ matrices

We recall the fundamental relation that characterizes the Dirac gamma matrices, in particular the anti-commutator of any two gamma matrices is

$$\{\gamma^\mu, \gamma^\nu\} = 2\eta^{\mu\nu} ,$$

from which, for example,

$$\left\{\gamma^0, \gamma^j\right\} = 0 ,$$

or, similarly,

$$\gamma^0\gamma^j = -\gamma^j\gamma^0 .$$

The gamma matrices satisfy also the following relations

$$(\gamma^0)^2 = 1 ,$$
$$(\gamma^j)^2 = -1 ,$$

while the Hermitian conjugated of the gamma matrices are

$$\gamma^{\mu\dagger} = \gamma^0 \gamma^\mu \gamma^0 \,,$$

with

$$(\gamma^0 \gamma^\mu)^\dagger = \gamma^{\mu\dagger} \gamma^0 \,,$$

from which

$$\gamma^{j\dagger} = \gamma^0 \gamma^j \gamma^0 = -\gamma^0 \gamma^0 \gamma^j = -\gamma^j \,,$$

this means that the gamma matrices with spatial index are anti-hermitian, while

$$\gamma^{0\dagger} = \gamma^0 \,,$$

so that the gamma matrix with index 0 is hermitian.

30.1 Traces of γ matrices

We compute the trace of a generic Dirac matrix with spatial index j

$$\begin{aligned}
\text{Tr}(\gamma^j) &= \text{Tr}(\gamma^j \cdot 1) = \text{Tr}(\gamma^j \gamma^0 \gamma^0) \\
&= \text{Tr}(\gamma^0 \gamma^j \gamma^0) \,,
\end{aligned}$$

where in the last step we exploited the property of the trace of a product of matrices to be invariant for cyclic permutations of the factors, in fact, for example, for the product of three generic matrices A, B, C we have

$$\mathrm{Tr}(ABC) = \mathrm{Tr}(BCA) = \mathrm{Tr}(CAB) \,.$$

We use now, in the last term of the previous equation, the property

$$\gamma^0 \gamma^j = -\gamma^j \gamma^0 \,,$$

from which

$$\mathrm{Tr}(\gamma^j) = -\mathrm{Tr}(\gamma^0 \gamma^0 \gamma^j) = -\mathrm{Tr}(\gamma^j) \,,$$

i.e.

$$2\,\mathrm{Tr}(\gamma^j) = 0 \,, \quad \mathrm{Tr}(\gamma^j) = 0 \,.$$

We can calculate, with similar passages, the trace of the Dirac matrix with index 0

$$\begin{aligned}
\mathrm{Tr}(\gamma^0) &= -\mathrm{Tr}\Big(\gamma^0 \cdot (-1)\Big) = -\mathrm{Tr}(\gamma^0 \gamma^j \gamma^j) \\
&= -\mathrm{Tr}(\gamma^0 \gamma^j \gamma^0) \,,
\end{aligned}$$

from which

$$\begin{aligned}
\mathrm{Tr}(\gamma^0) &= -\mathrm{Tr}\Big((-\gamma^j \gamma^0)\gamma^0\Big) = \mathrm{Tr}(\gamma^j \gamma^0 \gamma^0) \\
&= \mathrm{Tr}(\gamma^j) = 0 \,.
\end{aligned}$$

We have shown, using only the fundamental property, that the trace of any Dirac matrix is null

$$\mathrm{Tr}(\gamma^\mu) = 0 \, .$$

30.2 The γ^5 matrix

We introduce a new matrix, defined as

$$\gamma^5 = i\gamma^0\gamma^1\gamma^2\gamma^3 \, ,$$

and, remembering that

$$\gamma^0\gamma^j = -\gamma^j\gamma^0 \, ,$$

we explicitly calculate the anti-commutator between the γ^5 matrix and the other Dirac γ matrices

$$
\begin{aligned}
\{\gamma^0, \gamma^5\} &= i\gamma^0\gamma^0\gamma^1\gamma^2\gamma^3 + i\gamma^0\gamma^1\gamma^2\gamma^3\gamma^0 \\
&= i\gamma^1\gamma^2\gamma^3 - i\gamma^1\gamma^0\gamma^2\gamma^3\gamma^0 \\
&= i\gamma^1\gamma^2\gamma^3 + i\gamma^1\gamma^2\gamma^0\gamma^3\gamma^0 \\
&= i\gamma^1\gamma^2\gamma^3 - i\gamma^1\gamma^2\gamma^3\gamma^0\gamma^0 \\
&= i\gamma^1\gamma^2\gamma^3 - i\gamma^1\gamma^2\gamma^3 = 0 \, ,
\end{aligned}
$$

in the same way

$$\begin{aligned}
\{\gamma^1, \gamma^5\} &= i\gamma^1\gamma^0\gamma^1\gamma^2\gamma^3 + i\gamma^0\gamma^1\gamma^2\gamma^3\gamma^1 \\
&= -i\gamma^0\gamma^1\gamma^1\gamma^2\gamma^3 - i\gamma^0\gamma^2\gamma^1\gamma^3\gamma^1 \\
&= i\gamma^0\gamma^2\gamma^3 + i\gamma^0\gamma^2\gamma^3\gamma^1\gamma^1 \\
&= i\gamma^0\gamma^2\gamma^3 - i\gamma^0\gamma^2\gamma^3 = 0
\end{aligned}$$

and

$$\begin{aligned}
\{\gamma^2, \gamma^5\} &= i\gamma^2\gamma^0\gamma^1\gamma^2\gamma^3 + i\gamma^0\gamma^1\gamma^2\gamma^3\gamma^2 \\
&= -i\gamma^0\gamma^2\gamma^1\gamma^2\gamma^3 - i\gamma^0\gamma^1\gamma^3\gamma^2\gamma^2 \\
&= i\gamma^0\gamma^1\gamma^2\gamma^2\gamma^3 + i\gamma^0\gamma^1\gamma^3 \\
&= -i\gamma^0\gamma^1\gamma^3 + i\gamma^0\gamma^1\gamma^3 = 0 \,,
\end{aligned}$$

finally

$$\begin{aligned}
\{\gamma^3, \gamma^5\} &= i\gamma^3\gamma^0\gamma^1\gamma^2\gamma^3 + i\gamma^0\gamma^1\gamma^2\gamma^3\gamma^3 \\
&= -i\gamma^0\gamma^3\gamma^1\gamma^2\gamma^3 - i\gamma^0\gamma^1\gamma^2 \\
&= i\gamma^0\gamma^1\gamma^3\gamma^2\gamma^3 - i\gamma^0\gamma^1\gamma^2 \\
&= -i\gamma^0\gamma^1\gamma^2\gamma^3\gamma^3 - i\gamma^0\gamma^1\gamma^2 \\
&= i\gamma^0\gamma^1\gamma^2 - i\gamma^0\gamma^1\gamma^2 = 0 \,.
\end{aligned}$$

These results can be written in a compact way as

$$\{\gamma^\mu, \gamma^5\} = 0 \,.$$

Concerning the square of the γ^5 matrix we can write

$$
\begin{aligned}
(\gamma^5)^2 &= \gamma^5\gamma^5 = -\gamma^0\gamma^1\gamma^2\gamma^3\gamma^0\gamma^1\gamma^2\gamma^3 \\
&= -\gamma^1\gamma^0\gamma^2\gamma^0\gamma^3\gamma^1\gamma^2\gamma^3 = \gamma^1\gamma^2\gamma^3\gamma^1\gamma^2\gamma^3 \\
&= \gamma^2\gamma^1\gamma^1\gamma^3\gamma^2\gamma^3 = -\gamma^2\gamma^3\gamma^2\gamma^3 \\
&= \gamma^3\gamma^2\gamma^2\gamma^3 = -\gamma^3\gamma^3 = 1\,,
\end{aligned}
$$

i.e.

$$
(\gamma^5)^2 = 1\,.
$$

Recalling the expression

$$
\gamma^{\mu\dagger} = \gamma^0\gamma^\mu\gamma^0\,,
$$

we calculate

$$
\begin{aligned}
(\gamma^5)^\dagger &= -i(\gamma^0\gamma^1\gamma^2\gamma^3)^\dagger = -i\gamma^{3\dagger}\gamma^{2\dagger}\gamma^{1\dagger}\gamma^{0\dagger} \\
&= -i\gamma^0\gamma^3\gamma^0\gamma^0\gamma^2\gamma^0\gamma^0\gamma^1\gamma^0\gamma^0 = -i\gamma^0\gamma^3\gamma^2\gamma^1 \\
&= i\gamma^0\gamma^3\gamma^1\gamma^2 = i\gamma^0\gamma^1\gamma^3\gamma^2 = -i\gamma^0\gamma^1\gamma^2\gamma^3\,,
\end{aligned}
$$

or

$$
(\gamma^5)^\dagger = -\gamma^5\,.
$$

30.3 Traces of products of γ matrices

We calculate in this section the trace of the product of an odd number of gamma matrices

$$\text{Tr}(\gamma^{\mu_1}\gamma^{\mu_2}\cdots\gamma^{\mu_{2n+1}}) = \text{Tr}(\gamma^5\gamma^5\gamma^{\mu_1}\gamma^{\mu_2}\cdots\gamma^{\mu_{2n+1}})$$
$$= (-1)^{2n+1}\text{Tr}(\gamma^5\gamma^{\mu_1}\gamma^{\mu_2}\cdots\gamma^{\mu_{2n+1}}\gamma^5)$$
$$= -\text{Tr}(\gamma^5\gamma^{\mu_1}\gamma^{\mu_2}\cdots\gamma^{\mu_{2n+1}}\gamma^5)\,,$$

where we have used the anti-commutator relation to bring the second γ^5 matrix from its position to the last one, anti-commuting with the $2n+1$ gamma matrices, by introducing, at each exchange, a relative sign. By cyclically permuting the last γ^5 matrix in the first position, leaving the trace unchanged, we obtain

$$\text{Tr}(\gamma^{\mu_1}\gamma^{\mu_2}\cdots\gamma^{\mu_{2n+1}}) = -\text{Tr}(\gamma^5\gamma^5\gamma^{\mu_1}\gamma^{\mu_2}\cdots\gamma^{\mu_{2n+1}})$$
$$= -\text{Tr}(\gamma^{\mu_1}\gamma^{\mu_2}\cdots\gamma^{\mu_{2n+1}})\,,$$

from which

$$2\,\text{Tr}(\gamma^{\mu_1}\gamma^{\mu_2}\cdots\gamma^{\mu_{2n+1}}) = 0$$

and hence

$$\text{Tr}(\gamma^{\mu_1}\gamma^{\mu_2}\cdots\gamma^{\mu_{2n+1}}) = 0\,.$$

We conclude that the trace of the product of an odd number of Dirac gamma matrices is always null.

The trace of an even number of gamma matrices could be, in general, not null. Let's start by calculating the trace of the product of two Dirac gamma matrices

$$
\begin{aligned}
\mathrm{Tr}(\gamma^\mu \gamma^\nu) &= \mathrm{Tr}(2\eta^{\mu\nu} - \gamma^\nu \gamma^\mu) \\
&= \mathrm{Tr}(2\eta^{\mu\nu}) - \mathrm{Tr}(\gamma^\nu \gamma^\mu) \\
&= 2\eta^{\mu\nu}\mathrm{Tr}(I) - \mathrm{Tr}(\gamma^\mu \gamma^\nu) \\
&= 8\eta^{\mu\nu} - \mathrm{Tr}(\gamma^\mu \gamma^\nu) \,,
\end{aligned}
$$

from which

$$
2\,\mathrm{Tr}(\gamma^\mu \gamma^\nu) = 8\eta^{\mu\nu}
$$

and hence

$$
\mathrm{Tr}(\gamma^\mu \gamma^\nu) = 4\eta^{\mu\nu} \,,
$$

where we have used the expression

$$
\gamma^\mu \gamma^\nu = 2\eta^{\mu\nu} - \gamma^\nu \gamma^\mu \,.
$$

We compute the trace of the γ^5 matrix

$$
\begin{aligned}
\mathrm{Tr}(\gamma^5) &= i\mathrm{Tr}(\gamma^0 \gamma^1 \gamma^2 \gamma^3) = -i\mathrm{Tr}(\gamma^1 \gamma^0 \gamma^2 \gamma^3) \\
&= i\mathrm{Tr}(\gamma^1 \gamma^2 \gamma^0 \gamma^3) - i\mathrm{Tr}(\gamma^1 \gamma^2 \gamma^3 \gamma^0) \\
&= -i\mathrm{Tr}(\gamma^0 \gamma^1 \gamma^2 \gamma^3) = -\mathrm{Tr}(\gamma^5) \,,
\end{aligned}
$$

from which

$$
\mathrm{Tr}(\gamma^5) = 0 \,.
$$

Consider the trace of the product of 4 generic Dirac gamma matrices

$$
\begin{aligned}
\text{Tr}(\gamma^\mu\gamma^\nu\gamma^\alpha\gamma^\beta) &= \text{Tr}\Big((2\eta^{\mu\nu} - \gamma^\nu\gamma^\mu)\gamma^\alpha\gamma^\beta\Big) \\
&= \text{Tr}\Big(2\eta^{\mu\nu}\gamma^\alpha\gamma^\beta - \gamma^\nu(2\eta^{\mu\alpha} \\
&\quad - \gamma^\alpha\gamma^\mu)\gamma^\beta\Big) = \text{Tr}(2\eta^{\mu\nu}\gamma^\alpha\gamma^\beta \\
&\quad - 2\eta^{\mu\alpha}\gamma^\nu\gamma^\beta + \gamma^\nu\gamma^\alpha\gamma^\mu\gamma^\beta)\,,
\end{aligned}
$$

it follows that

$$
\begin{aligned}
\text{Tr}(\gamma^\mu\gamma^\nu\gamma^\alpha\gamma^\beta) &= \text{Tr}\Big(2\eta^{\mu\nu}\gamma^\alpha\gamma^\beta - 2\eta^{\mu\alpha}\gamma^\nu\gamma^\beta \\
&\quad + \gamma^\nu\gamma^\alpha(2\eta^{\mu\beta} - \gamma^\beta\gamma^\mu)\Big) \\
&= \text{Tr}(2\eta^{\mu\nu}\gamma^\alpha\gamma^\beta - 2\eta^{\mu\alpha}\gamma^\nu\gamma^\beta \\
&\quad + 2\eta^{\mu\beta}\gamma^\nu\gamma^\alpha - \gamma^\nu\gamma^\alpha\gamma^\beta\gamma^\mu) \\
&= 2\eta^{\mu\nu}\text{Tr}(\gamma^\alpha\gamma^\beta) \\
&\quad - 2\eta^{\mu\alpha}\text{Tr}(\gamma^\nu\gamma^\beta) + 2\eta^{\mu\beta}\text{Tr}(\gamma^\nu\gamma^\alpha) \\
&\quad - \text{Tr}(\gamma^\nu\gamma^\alpha\gamma^\beta\gamma^\mu)\,.
\end{aligned}
$$

The matrices in the last trace can be cyclically exchanged obtaining the same order as those in the first member trace, i.e.

$$
\text{Tr}(\gamma^\mu\gamma^\nu\gamma^\alpha\gamma^\beta) = \text{Tr}(\gamma^\nu\gamma^\alpha\gamma^\beta\gamma^\mu)\,,
$$

so that

$$2\,\mathrm{Tr}(\gamma^\mu\gamma^\nu\gamma^\alpha\gamma^\beta) = 2\eta^{\mu\nu}\mathrm{Tr}(\gamma^\alpha\gamma^\beta) - 2\eta^{\mu\alpha}\mathrm{Tr}(\gamma^\nu\gamma^\beta)$$
$$+ 2\eta^{\mu\beta}\mathrm{Tr}(\gamma^\nu\gamma^\alpha)\,,$$

from which

$$\mathrm{Tr}(\gamma^\mu\gamma^\nu\gamma^\alpha\gamma^\beta) = 4\eta^{\mu\nu}\eta^{\alpha\beta} - 4\eta^{\mu\alpha}\eta^{\nu\beta} + 4\eta^{\mu\beta}\eta^{\nu\alpha}$$
$$= 4\left(\eta^{\mu\nu}\eta^{\alpha\beta} - \eta^{\mu\alpha}\eta^{\nu\beta} + \eta^{\mu\beta}\eta^{\nu\alpha}\right).$$

To obtain the trace of a product of an even number greater than 4 of gamma matrices, one can proceed in the same way as done before, using many times the relation

$$\gamma^\mu\gamma^\nu = 2\eta^{\mu\nu} - \gamma^\nu\gamma^\mu\,.$$

30.4 Dirac representation of γ matrices

Always starting from the anti-commutation relation for the gamma matrices, it is clear that they have an even dimension with a minimum value of 4, because there are only three matrices of size 2×2 mutually anti-commuting, i.e. the Pauli matrices σ_1, σ_2 and σ_3, which can be written in the representation where σ_3 is diagonal as

$$\sigma_1 = \begin{pmatrix} 0 & 1 \\ 1 & 0 \end{pmatrix},$$

$$\sigma_2 = \begin{pmatrix} 0 & -i \\ i & 0 \end{pmatrix} ,$$

$$\sigma_3 = \begin{pmatrix} 1 & 0 \\ 0 & -1 \end{pmatrix} .$$

The Pauli matrices satisfy the property

$$\sigma_j^2 = I$$

and their determinant and trace are

$$\det \sigma_j = -1 , \quad \mathrm{Tr} \sigma_j = 0 .$$

The Dirac representation of the gamma matrices is that where γ^0 is diagonal. In this representation we have

$$\gamma^0 = \begin{pmatrix} 1 & 0 & 0 & 0 \\ 0 & 1 & 0 & 0 \\ 0 & 0 & -1 & 0 \\ 0 & 0 & 0 & -1 \end{pmatrix} ,$$

$$\gamma^1 = \begin{pmatrix} 0 & 0 & 0 & 1 \\ 0 & 0 & 1 & 0 \\ 0 & -1 & 0 & 0 \\ -1 & 0 & 0 & 0 \end{pmatrix} ,$$

$$\gamma^2 = \begin{pmatrix} 0 & 0 & 0 & -i \\ 0 & 0 & i & 0 \\ 0 & i & 0 & 0 \\ -i & 0 & 0 & 0 \end{pmatrix},$$

$$\gamma^3 = \begin{pmatrix} 0 & 0 & 1 & 0 \\ 0 & 0 & 0 & -1 \\ -1 & 0 & 0 & 0 \\ 0 & 1 & 0 & 0 \end{pmatrix},$$

$$\gamma^5 = \begin{pmatrix} 0 & 0 & 1 & 0 \\ 0 & 0 & 0 & 1 \\ 1 & 0 & 0 & 0 \\ 0 & 1 & 0 & 0 \end{pmatrix}.$$

These can be written in compact form, using Pauli matrices, such as

$$\gamma^0 = \begin{pmatrix} I & 0 \\ 0 & -I \end{pmatrix},$$

$$\gamma^j = \begin{pmatrix} 0 & \sigma^j \\ -\sigma^j & 0 \end{pmatrix},$$

$$\gamma^5 = \begin{pmatrix} 0 & -I \\ -I & 0 \end{pmatrix} .$$

Chapter 31

Covariance of the Dirac equation

We write now the Dirac equation in two inertial reference frames, related by a Lorentz transformation, computed by the Λ matrix. The two equations are

$$(i\gamma^\mu \partial_\mu - m)\psi(x) = 0$$

and

$$(i\gamma^\mu \partial'_\mu - m)\psi'(x') = 0\,,$$

with

$$\partial'_\mu = \Lambda_\mu^{\;\nu} \partial_\nu\,.$$

Suppose that

$$\psi'(x') = S(\Lambda)\psi(x)\,,$$

the second equation becomes

$$(i\gamma^\mu \Lambda_\mu{}^\nu \partial_\nu - m)S(\Lambda)\psi(x) = 0$$

and, multiplying by the inverse of the matrix S from the left, we obtain

$$\left(iS^{-1}(\Lambda)\gamma^\mu S(\Lambda)\Lambda_\mu{}^\nu \partial_\nu - m\right)\psi(x) = 0\,.$$

In order to have the same form for the two equations in both reference frames it is necessary that

$$S^{-1}(\Lambda)\gamma^\mu S(\Lambda)\Lambda_\mu{}^\nu = \gamma^\nu\,,$$

that we can write as

$$S^{-1}(\Lambda)\gamma^\mu S(\Lambda) = \Lambda^\mu{}_\nu \gamma^\nu\,.$$

We can also define

$$\gamma'^\mu = \Lambda^\mu{}_\nu \gamma^\nu$$

and it can be verified that these matrices preserve the anti-commutation relation

$$\{\gamma'^\mu, \gamma'^\nu\} = 2\eta^{\mu\nu}\,.$$

Chapter 32

Dirac Hamiltonian

The Dirac equation

$$(i\gamma^\mu \partial_\mu - m)\psi = 0\,,$$

can be explicitly written as

$$(i\gamma^0 \partial_0 + i\vec{\gamma} \cdot \vec{\nabla} - m)\psi = 0\,,$$

where

$$\vec{\gamma} = (\gamma^1, \gamma^2, \gamma^3)\,,$$

from which

$$i\gamma^0 \frac{\partial \psi}{\partial t} = \left(\vec{\gamma}(-i\vec{\nabla}) + m\right)\psi\,.$$

Using the momentum operator in quantum mechanics

$$\hat{\vec{p}} = -i\vec{\nabla}\,,$$

the previous equation becomes

$$i\gamma^0 \frac{\partial \psi}{\partial t} = \left(\vec{\gamma} \cdot \vec{p} + m \right) \psi \,.$$

From the property

$$(\gamma^0)^2 = 1 \,,$$

we have, multiplying the equation by γ^0 from the left,

$$i\frac{\partial \psi}{\partial t} = \left(\gamma^0 \vec{\gamma} \cdot \vec{p} + m\gamma^0 \right) \psi \,.$$

We define the matrices

$$\beta = \gamma^0 \,, \quad \vec{\alpha} = \gamma^0 \vec{\gamma} \,,$$

so that the Dirac equation becomes

$$i\frac{\partial \psi}{\partial t} = \left(\vec{\alpha} \cdot \vec{p} + m\beta \right) \psi \,.$$

If we write the equation in the form

$$i\frac{\partial \psi}{\partial t} = H_D \, \psi \,,$$

we recognize the Dirac Hamiltonian

$$H_D = \vec{\alpha} \cdot \vec{p} + m\beta \,.$$

Chapter 33

Dirac Lagrangian

Changing the point of view, passing from a single particle approach to a field theory view, we can write the Dirac Lagrangian density

$$\mathcal{L}_D = \overline{\psi}(i\gamma^\mu \partial_\mu - m)\psi \,,$$

whose Euler-Lagrange equations for the fields ψ and $\overline{\psi}$, which must be treated as independent fields, are

$$\partial_\mu \left(\frac{\partial \mathcal{L}}{\partial \partial_\mu \psi} \right) = \frac{\partial \mathcal{L}}{\partial \psi}$$

and

$$\partial_\mu \left(\frac{\partial \mathcal{L}}{\partial \partial_\mu \overline{\psi}} \right) = \frac{\partial \mathcal{L}}{\partial \overline{\psi}} \,.$$

We calculate the four quantities

$$\frac{\partial \mathcal{L}}{\partial \psi} = -m\overline{\psi} \,,$$

$$\frac{\partial \mathcal{L}}{\partial \partial_\mu \psi} = i\overline{\psi}\gamma^\mu \,,$$

$$\frac{\partial \mathcal{L}}{\partial \overline{\psi}} = (i\gamma^\mu \partial_\mu - m)\psi$$

and

$$\frac{\partial \mathcal{L}}{\partial \partial_\mu \overline{\psi}} = 0 \,,$$

from which we obtain the following two equations, for the field ψ

$$(i\gamma^\mu \partial_\mu - m)\psi = 0$$

and for the field $\overline{\psi}$

$$i\overline{\psi}\gamma^\mu + m\overline{\psi} = 0 \,,$$

which represent the Dirac equations for the two fields.

Chapter 34

Free solutions

We start from the Dirac equation

$$(i\gamma^\mu\partial_\mu - m)\psi(x) = 0$$

and search for a solution that contains the free particle information (plane wave) and is eigen-function of the four-momentum with positive energy. We can write it in the form

$$\phi^+(x) = e^{-ip\cdot x}u(\vec{p})\,, \quad p^0 > 0\,,$$

where the term $u(\vec{p})$ determines its spinorial properties. By substituting in the Dirac equation

$$(i\gamma^\mu\partial_\mu - m)\left(e^{-ip\cdot x}u(\vec{p})\right) = 0\,,$$

we obtain

$$i\gamma^\mu \partial_\mu \left(e^{-ip_\nu x^\nu} u(\vec{p})\right) - me^{-ip\cdot x} u(\vec{p}) = 0\,,$$
$$i\gamma^\mu \partial_\mu \left(e^{-ip_\nu x^\nu}\right) u(\vec{p}) - me^{-ip\cdot x} u(\vec{p}) = 0\,,$$
$$i\gamma^\mu \left(-ip_\mu e^{-ip\cdot x}\right) u(\vec{p}) - me^{-ip\cdot x} u(\vec{p}) = 0\,,$$

finally

$$(\gamma^\mu p_\mu - m)u(\vec{p}) = 0\,.$$

34.1 Solutions in the rest frame

In the reference frame of a non-zero mass particle, the equation becomes

$$(\gamma^0 p_0 - m)u(\vec{0}) = 0\,,$$

or

$$(\gamma^0 m - m)u(\vec{0}) = 0\,,$$

or, again,

$$(\gamma^0 - 1)u(\vec{0}) = 0\,,$$

having used the four-momentum

$$p^\mu = (m, \vec{0})\,.$$

By calculating

$$\gamma^0 - 1 = \begin{pmatrix} 1 & 0 & 0 & 0 \\ 0 & 1 & 0 & 0 \\ 0 & 0 & -1 & 0 \\ 0 & 0 & 0 & -1 \end{pmatrix} - \begin{pmatrix} 1 & 0 & 0 & 0 \\ 0 & 1 & 0 & 0 \\ 0 & 0 & 1 & 0 \\ 0 & 0 & 0 & 1 \end{pmatrix},$$

we obtain

$$\gamma^0 - 1 = \begin{pmatrix} 0 & 0 & 0 & 0 \\ 0 & 0 & 0 & 0 \\ 0 & 0 & -2 & 0 \\ 0 & 0 & 0 & -2 \end{pmatrix}$$

and we have the matrix equation

$$\begin{pmatrix} 0 & 0 & 0 & 0 \\ 0 & 0 & 0 & 0 \\ 0 & 0 & -2 & 0 \\ 0 & 0 & 0 & -2 \end{pmatrix} \begin{pmatrix} u^0 \\ u^1 \\ u^2 \\ u^3 \end{pmatrix} = \begin{pmatrix} 0 \\ 0 \\ 0 \\ 0 \end{pmatrix},$$

equivalent to

$$\begin{pmatrix} 0 & 0 & 0 & 0 \\ 0 & 0 & 0 & 0 \\ 0 & 0 & 1 & 0 \\ 0 & 0 & 0 & 1 \end{pmatrix} \begin{pmatrix} u^0 \\ u^1 \\ u^2 \\ u^3 \end{pmatrix} = \begin{pmatrix} 0 \\ 0 \\ 0 \\ 0 \end{pmatrix},$$

having defined

$$u(\vec{0}) = \begin{pmatrix} u^0 \\ u^1 \\ u^2 \\ u^3 \end{pmatrix} \, .$$

We obtain

$$u^2 = u^3 = 0 \, ,$$

hence the two independent solutions

$$u_{(1)}(\vec{0}) = \begin{pmatrix} 1 \\ 0 \\ 0 \\ 0 \end{pmatrix}$$

and

$$u_{(2)}(\vec{0}) = \begin{pmatrix} 0 \\ 1 \\ 0 \\ 0 \end{pmatrix}$$

which can be interpreted as the two projections of the spin of the particle (in its reference frame).

The study of the negative energy solutions is completely

analogous and leads to the following results

$$\phi^-(x) = e^{ip \cdot x} v(\vec{p}), \quad p^0 > 0,$$

the equation becomes

$$(\gamma^\mu p_\mu + m)v(\vec{p}) = 0,$$

from which, in the reference frame of the non-zero mass particle where

$$p^\mu = (m, \vec{0}),$$

we have

$$(\gamma^0 p_0 + m)v(\vec{0}) = 0,$$

and

$$(\gamma^0 + 1)v(\vec{0}) = 0.$$

We calculate

$$\gamma^0 + 1 = \begin{pmatrix} 1 & 0 & 0 & 0 \\ 0 & 1 & 0 & 0 \\ 0 & 0 & -1 & 0 \\ 0 & 0 & 0 & -1 \end{pmatrix} + \begin{pmatrix} 1 & 0 & 0 & 0 \\ 0 & 1 & 0 & 0 \\ 0 & 0 & 1 & 0 \\ 0 & 0 & 0 & 1 \end{pmatrix},$$

i.e.

$$\gamma^0 + 1 = \begin{pmatrix} 2 & 0 & 0 & 0 \\ 0 & 2 & 0 & 0 \\ 0 & 0 & 0 & 0 \\ 0 & 0 & 0 & 0 \end{pmatrix},$$

that becomes

$$\begin{pmatrix} 1 & 0 & 0 & 0 \\ 0 & 1 & 0 & 0 \\ 0 & 0 & 0 & 0 \\ 0 & 0 & 0 & 0 \end{pmatrix} \begin{pmatrix} v^0 \\ v^1 \\ v^2 \\ v^3 \end{pmatrix} = \begin{pmatrix} 0 \\ 0 \\ 0 \\ 0 \end{pmatrix},$$

where

$$v(\vec{0}) = \begin{pmatrix} v^0 \\ v^1 \\ v^2 \\ v^3 \end{pmatrix}.$$

The independent solutions are

$$v_{(1)}(\vec{0}) = \begin{pmatrix} 0 \\ 0 \\ 1 \\ 0 \end{pmatrix}$$

and

$$v_{(2)}(\vec{0}) = \begin{pmatrix} 0 \\ 0 \\ 0 \\ 1 \end{pmatrix}.$$

34.2 Generic solutions

In order to derive the expression of the spinors in any reference frame, we start from the equation

$$(\gamma^\mu p_\mu - m)u(\vec{p}) = 0.$$

We observe that the product

$$(\gamma^\mu p_\mu + m)(\gamma^\nu p_\nu - m)$$

can be written also as

$$
\begin{aligned}
(\gamma^\mu p_\mu + m)(\gamma^\nu p_\nu - m) &= \gamma^\mu p_\mu \gamma^\nu p_\nu \\
&\quad + m(\gamma^\nu p_\nu - \gamma^\mu p_\mu) - m^2 \\
&= \gamma^\mu \gamma^\nu p_\mu p_\nu - m^2 \\
&= \frac{1}{2}(\gamma^\mu \gamma^\nu + \gamma^\mu \gamma^\nu + \gamma^\nu \gamma^\mu \\
&\quad - \gamma^\nu \gamma^\mu)p_\mu p_\nu - m^2 \\
&= \frac{1}{2}\Big(\{\gamma^\mu, \gamma^\nu\} + [\gamma^\mu, \gamma^\nu]\Big) \\
&\quad \cdot\ p_\mu p_\nu - m^2 = \frac{1}{2}\{\gamma^\mu, \gamma^\nu\} \\
&\quad \cdot\ p_\mu p_\nu - m^2 = \eta^{\mu\nu} p_\mu p_\nu \\
&\quad - m^2 = p^\mu p_\mu - m^2 \\
&= p^2 - m^2 = 0 \,.
\end{aligned}
$$

We can therefore write

$$
(\gamma^\mu p_\mu - m)\Big[(\gamma^\nu p_\nu + m)u_{1,2}(\vec{0})\Big] = 0\,,
$$

from which

$$
u_{(1,2)}(\vec{p}) = N(\gamma^\nu p_\nu + m)u_{(1,2)}(\vec{0})\,,
$$

where N is a constant to be fixed.

We calculate, using the explicit form of the gamma matrices

$$
\gamma^0 = \begin{pmatrix} I & 0 \\ 0 & -I \end{pmatrix}\,,
$$

$$\gamma^j = \begin{pmatrix} 0 & \sigma^j \\ -\sigma^j & 0 \end{pmatrix},$$

$$\gamma^5 = \begin{pmatrix} 0 & -I \\ -I & 0 \end{pmatrix},$$

knowing that the momentum can be written as

$$p^\mu = (E, \vec{p}),$$

where E is the energy of the particle. We have

$$u_{(1,2)}(\vec{p}) = N(\gamma^\nu p_\nu + m)u_{(1,2)}(\vec{0}),$$

from which

$$
\begin{aligned}
u_{(1,2)}(\vec{p}) &= N\left[\begin{pmatrix} I & 0 \\ 0 & -I \end{pmatrix} E - \begin{pmatrix} 0 & \vec{\sigma} \\ -\vec{\sigma} & 0 \end{pmatrix} \vec{p} \right. \\
&\quad + \left. m \begin{pmatrix} I & 0 \\ 0 & I \end{pmatrix} \right] u_{(1,2)}(\vec{0}) \\
&= N \begin{pmatrix} E + m & -\vec{\sigma} \cdot \vec{p} \\ \vec{\sigma} \cdot \vec{p} & -E + m \end{pmatrix} u_{(1,2)}(\vec{0})
\end{aligned}
$$

and hence

$$u_{(1)}(\vec{p}) = N \begin{pmatrix} E + m \\ 0 \\ \vec{\sigma} \cdot \vec{p} \\ 0 \end{pmatrix},$$

$$u_{(2)}(\vec{p}) = N \begin{pmatrix} 0 \\ E+m \\ 0 \\ \vec{\sigma} \cdot \vec{p} \end{pmatrix}.$$

The constant N can be determined from the condition

$$\overline{u}_{(a)}(\vec{p})\overline{u}_{(b)}(\vec{p}) = \delta_{a,b}, \quad a,b = 1,2,$$

obtaining

$$N = \frac{1}{\sqrt{2(E+m)m}},$$

from which, finally

$$u_{(1)}(\vec{p}) = \frac{1}{\sqrt{2(E+m)m}} \begin{pmatrix} E+m \\ 0 \\ \vec{\sigma} \cdot \vec{p} \\ 0 \end{pmatrix}$$

and

$$u_{(2)}(\vec{p}) = \frac{1}{\sqrt{2(E+m)m}} \begin{pmatrix} 0 \\ E+m \\ 0 \\ \vec{\sigma} \cdot \vec{p} \end{pmatrix}.$$

The equation that relates the spinors in any reference frame with those in the particle reference frame is

$$u_{(1,2)}(\vec{p}) = \frac{\gamma^\nu p_\nu + m}{\sqrt{2(E+m)m}} u_{(1,2)}(\vec{0})$$

and it is independent of the representation used for the gamma matrices. Similarly we obtain the equation for the negative energy solutions, where the condition for the constant N can be written as

$$\bar{v}_{(a)}(\vec{p})\bar{v}_{(b)}(\vec{p}) = -\delta_{a,b} \quad a,b = 1,2\,,$$

and the result is

$$v_{(1,2)}(\vec{p}) = \frac{-\gamma^\nu p_\nu + m}{\sqrt{2(E+m)m}} v_{(1,2)}(\vec{0})\,.$$

The complete solutions are

$$\phi^+(x) = e^{-ip\cdot x} u(\vec{p})\,, \quad p^0 > 0\,,$$

and

$$\phi^-(x) = e^{ip\cdot x} v(\vec{p})\,, \quad p^0 > 0\,,$$

that, after normalization, form a complete orthonormal system and therefore the more general solution (wave packet) can be obtained as their linear combination.

Part V

Relativity, decays and electromagnetic fields

Introduction

After introducing the metric in classical Euclidean space, we move to Minkowski's concept of four-vectors in space--time, dealing with topics of restricted relativity, such as Lorentz transformations and Lorentz invariants. We obtain the relativistic expressions of the total energy, the energy at rest and the kinetic energy of a free particle, showing also their non-relativistic limits. Subsequently, we analyze the decay of particles, in particular the decay of the muon. We introduce the electromagnetic field tensor, with calculations of the electric and magnetic field vectors in different inertial reference frames. Finally, we show the Maxwell's equations, both in differential and covariant form, showing how to obtain the equation of electromagnetic waves in vacuum.

Chapter 35

Euclidean space

35.1 Vectors and metrics

35.1.1 Vector

A vector in the three-dimensional euclidean space can be written, in terms of its components, as

$$\vec{v} = (v^1, v^2, v^3),$$

where the superscript distinguishes the coordinates and it does not represent an exponentiation. We have chosen to distinguish the components of a vector in euclidean space with superscript rather than subscript to immediately introduce the formalism of special relativity. In fact, when we talk about four-vector (four-component generalization of a vector) we will see that we must distinguish components with index at the top (superscript) and components with index at the bottom (subscript).

So we can write the i-th component as

$$v^i\,, \quad i = 1, 2, 3\,.$$

35.1.2 Scalar product

The scalar product between two vectors is written as

$$\vec{v} \cdot \vec{u} = v^1 u^1 + v^2 u^2 + v^3 u^3\,,$$

or

$$\vec{v} \cdot \vec{u} = \sum_{i=1}^{3} v^i u^i\,.$$

The scalar product of a vector with itself is

$$\vec{v}^2 = \vec{v} \cdot \vec{v} = (v^1)^2 + (v^2)^2 + (v^3)^2 = \sum_{i=1}^{3} (v^i)^2$$

and corresponds to the squared norm of the vector.

35.1.3 Metric

The formula for the scalar product of a vector with itself

$$\vec{v}^2 = \sum_{i=1}^{3} (v^i)^2 = \sum_{i=1}^{3} v^i v^i\,,$$

can be written also as a combination

$$\vec{v}^2 = \sum_{i=1}^{3} a_i v^i v^i \, ,$$

with constant coefficients equal to 1, i.e.

$$a_i = 1 \, , \quad \forall i \, .$$

A generalization of the latter formula is

$$\vec{v}^2 = \sum_{i=1}^{3} \sum_{j=1}^{3} \eta_{ij} v^i v^j \, ,$$

with

$$\eta_{ij} = \begin{cases} 1 & \text{if } i = j \\ 0 & \text{if } i \neq j \end{cases} \, ,$$

in fact with this choice the terms which include mixed products of components that do not appear in the scalar product are null.

The coefficients of the combination of the coordinates of the vector can be interpreted as the elements of a diagonal matrix (elements that are not null only if the indices are equal). This 3-dimensional matrix in the euclidean space takes the form

$$\eta_{ij} = \begin{pmatrix} 1 & 0 & 0 \\ 0 & 1 & 0 \\ 0 & 0 & 1 \end{pmatrix} = \text{diag}(1,1,1)$$

and we can write the square of a vector as a product between vectors and matrix in the following way

$$\vec{v}^2 = \begin{pmatrix} v^1 & v^2 & v^3 \end{pmatrix} \eta \begin{pmatrix} v^1 \\ v^2 \\ v^3 \end{pmatrix}$$

$$= \begin{pmatrix} v^1 & v^2 & v^3 \end{pmatrix} \begin{pmatrix} \eta_{11}v^1 + \eta_{12}v^2 + \eta_{13}v^3 \\ \eta_{21}v^1 + \eta_{22}v^2 + \eta_{23}v^3 \\ \eta_{31}v^1 + \eta_{32}v^2 + \eta_{33}v^3 \end{pmatrix}$$

$$= \begin{pmatrix} v^1 & v^2 & v^3 \end{pmatrix} \begin{pmatrix} \eta_{11}v^1 \\ \eta_{22}v^2 \\ \eta_{33}v^3 \end{pmatrix} = \begin{pmatrix} v^1 & v^2 & v^3 \end{pmatrix} \begin{pmatrix} v^1 \\ v^2 \\ v^3 \end{pmatrix}$$

$$= v^1 v^1 + v^2 v^2 + v^3 v^3 = (v^1)^2 + (v^2)^2 + (v^3)^2 \,.$$

The matrix just introduced is called "metric", or "metric tensor". A tensor is a multi-index object, where the number of indexes is called rank. In this way a vector is a tensor of rank 1 and a matrix is a tensor of rank 2.

35.1.4 Covariant and contravariant components

From the formula

$$\vec{v}^2 = \sum_{i=1}^{3} \sum_{j=1}^{3} \eta_{ij} v^i v^j \, ,$$

we define the "covariant" components of the vector, characterized by a low index, the following

$$v_i \equiv \sum_{j=1}^{3} \eta_{ij} v^j \, ,$$

while those with a high index, used so far, are called "contravariant" components. In this way the square of a vector can be written in the most compact form

$$\vec{v}^2 = \sum_{i=1}^{3} v_i v^i \, .$$

We introduce the so-called Einstein convention on repeated indices which states that in a product between elements with indexes, a sum on the repeated indices (one high and the other low) is always implied. In this way the covariant components are simply written as

$$v_i = \eta_{ij} v^j \, ,$$

where the sum on the repeated index j is implied.

The square of a vector is then written, using the Einstein convention,

$$\vec{v}^2 = v_i v^i = \eta_{ij} v^i v^j \,,$$

where we remember that at the second member a sum on the repeated index i is implied, while at the third member two sums on the repeated indices i and j are implied.

The two formulas that are used to switch between the covariant and contravariant components of a vector are the following

$$v_i = \eta_{ij} v^j \,,$$
$$v^i = \eta^{ij} v_j \,,$$

from which

$$\eta_{ij} = \eta^{ij} \,, \quad \forall i, j \,.$$

By observing the latest formulas, it can be seen that the metric "lowers" and "raises" the indices, that is, it gives the covariant components starting from the contravariant ones and vice versa.

Chapter 36

Minkowski space-time

36.1 Four-vectors and metrics

36.1.1 Four-vector

A 4-dimensional four-vector in Minkowski's space-time (3 spatial + 1 temporal) is written in the contravariant form

$$v^\mu = \left(v^0, v^1, v^2, v^3\right),$$

where the first component (the one with index 0) is the temporal component, while the other three are the spatial components.

Usually, to indicate a four-vector, is used a letter without the arrow above, notation that is reserved for three-component (spatial) vectors.

By convention it is assumed that the Latin indices $(i, j, k, \cdots)$ assume the values 1,2,3, while the Greek indices assume the values 0,1,2,3. This convention is particularly useful when using the Einstein convention to understand where the implied sum runs over.

36.1.2 Scalar and metric product

The square of a four-vector is written as

$$v^2 = v \cdot v = \eta_{\mu\nu} v^\mu v^\nu = v_\mu v^\mu \,.$$

Always pay attention to the context for expressions that can be misleading. In this case, for example, the term at first member represents "the square of the four-vector v" and not "the component with the contravariant index 2". The metric of the Minkowski space-time is represented by the diagonal matrix

$$\eta = \mathrm{diag}(1, -1, -1, -1) = \begin{pmatrix} 1 & 0 & 0 & 0 \\ 0 & -1 & 0 & 0 \\ 0 & 0 & -1 & 0 \\ 0 & 0 & 0 & -1 \end{pmatrix},$$

from which

$$\begin{aligned} v \cdot v &= \eta_{00}(v^0)^2 + \eta_{11}(v^1)^2 + \eta_{22}(v^2)^2 + \eta_{33}(v^3)^2 \\ &= (v^0)^2 - (v^1)^2 - (v^2)^2 - (v^3)^2 = (v^0)^2 - \vec{v}^2 \,. \end{aligned}$$

The covariant components of a vector are written as

$$v_\mu = \eta_{\mu\nu} v^\nu \,,$$

i.e.

$$v_\mu = \left(v^0, -v^1, -v^2, -v^3\right).$$

Note that in the case of Minkowski space-time the covariant temporal component coincides with the contravariant one, while the three covariant spatial components are the opposite of the corresponding contravariant ones.

36.1.3 Position four-vector

The main four-vector that characterizes an event in the Minkowski space-time is the position four-vector, defined as

$$s^\mu = \left(ct, x, y, z\right),$$

where c is the speed of light.

The covariant components are

$$s_\mu = \eta_{\mu\nu}s^\nu = \left(ct, -x, -y, -z\right).$$

Its square is

$$s^2 = c^2t^2 - \vec{s}^2\,,$$

with

$$\vec{s} = \left(x, y, z\right)$$

the position vector and

$$\vec{s}^2 = x^2 + y^2 + z^2 \,.$$

The infinitesimal interval that separates two events is written as

$$ds^2 = c^2 dt^2 - d\vec{x}^2 \,,$$

where it can observed that the minus sign represents the intrinsic difference between the 4-dimensional euclidean space with metric

$$\begin{pmatrix} 1 & 0 & 0 & 0 \\ 0 & 1 & 0 & 0 \\ 0 & 0 & 1 & 0 \\ 0 & 0 & 0 & 1 \end{pmatrix}$$

and the Minkowski 4-dimensional space-time with metric

$$\eta_{\mu\nu} = \begin{pmatrix} 1 & 0 & 0 & 0 \\ 0 & -1 & 0 & 0 \\ 0 & 0 & -1 & 0 \\ 0 & 0 & 0 & -1 \end{pmatrix} \,.$$

36.2 Lorentz invariants

36.2.1 Lorentz transformations

The coordinates of an event in two inertial reference frames S and S', the latter in motion with relative speed v with respect to S, along the abscissa axis, are related to each other through the so-called Lorentz transformations. In this particular case they are

$$\begin{cases} x' = \gamma(x - vt) \\ y' = y \\ z' = z \\ t' = \gamma\left(t - \frac{xv}{c^2}\right) \end{cases},$$

where γ is the Lorentz factor, given by

$$\gamma = \frac{1}{\sqrt{1 - \frac{v^2}{c^2}}}$$

with $\gamma \geq 1$. In general, the transformation of the coordinates of a four-vector between two inertial reference frames S and S' can be written as follows

$$x'^{\mu} = \Lambda^{\mu}_{\ \nu} x^{\nu},$$

where the Λ matrix, in the case discussed above, takes the form

$$
\Lambda^{\mu}{}_{\nu} = \begin{pmatrix} \gamma & -v\gamma/c & 0 & 0 \\ -v\gamma/c & \gamma & 0 & 0 \\ 0 & 0 & 1 & 0 \\ 0 & 0 & 0 & 1 \end{pmatrix} .
$$

All the quantities obtained with operations on the components of objects with indexes (such as four-vectors or tensors in general) where all the indexes are summed are called Lorentz scalars and have the properties of being Lorentz invariants, i.e. they assume the same value in all inertial reference frames. Examples of Lorentz scalars are expressions of the type

$$
a^{\mu}a_{\mu} , \quad A^{\mu\nu}A_{\mu\nu} , \quad v^{\mu}\epsilon_{\mu\nu\alpha\beta}C_{\nu\alpha}u^{\beta} ,
$$

where all the indices are summed, thanks to the Einstein convention.

On the other hand, the following quantities are not examples of Lorentz scalars and therefore their value depends on the inertial reference frame

$$
a^{\mu} , \quad A^{\mu\nu}b_{\mu} , \quad v^{\mu}\epsilon_{\mu\nu\alpha\beta}C_{\nu\alpha} .
$$

When you have the product of two quantities with a repeated index (and then they are implicitly summed thanks to the Einstein convention) it is said that the index is "contracted". A Lorentz scalar is obtained by contracting all the indices present in a product between components.

A typical Lorentz scalar is the square of a four-vector, for example the quantities

$$ds^2 = ds^\mu ds_\mu \,, \quad p^2 = p^\mu p_\mu$$

are Lorentz invariants and their values are independent from the reference frame. On the other hand, the single components of the four-vectors or tensors are not, in general, Lorentz invariants. For example the components

$$ds^\mu \,, \quad ds_\mu \,, \quad p^\mu \,, \quad p_\mu \,, \quad \mu = 0, 1, 2, 3$$

are not invariant under Lorentz transformations, in fact, for example, in the transition from a frame S to a frame S', we have

$$p^\mu \to p'^\mu = \Lambda^\mu_{\ \nu} p^\nu \,,$$

while instead

$$p^\mu p_\mu = p'^\mu p'_\mu \,.$$

From this equation we get

$$
\begin{aligned}
p'^{\mu}p'_{\mu} &= \eta_{\mu\nu}p'^{\mu}p'^{\nu} = \eta_{\mu\nu}\Lambda^{\mu}{}_{\alpha}p^{\alpha}\Lambda^{\nu}{}_{\beta}p^{\beta} \\
&= \eta_{\mu\nu}\Lambda^{\mu}{}_{\alpha}\Lambda^{\nu}{}_{\beta}p^{\alpha}p^{\beta} = \eta_{\alpha\beta}\Lambda^{\alpha}{}_{\mu}\Lambda^{\beta}{}_{\nu}p^{\mu}p^{\nu}\,,
\end{aligned}
$$

where in the last member we appropriately renamed the summed indices.

Knowing that

$$
p^{\mu}p_{\mu} = \eta_{\mu\nu}p^{\mu}p^{\nu}
$$

and that

$$
p^{\mu}p_{\mu} = p'^{\mu}p'_{\mu}\,,
$$

we have

$$
\eta_{\mu\nu}p^{\mu}p^{\nu} = \eta_{\alpha\beta}\Lambda^{\alpha}{}_{\mu}\Lambda^{\beta}{}_{\nu}p^{\mu}p^{\nu}\,,
$$

from which

$$
\eta_{\mu\nu} = \eta_{\alpha\beta}\Lambda^{\alpha}{}_{\mu}\Lambda^{\beta}{}_{\nu}\,, \quad \forall\mu,\nu\,.
$$

Other examples of Lorentz scalars are the scalar products between four-momenta of particles, for example the quantity

$$
p^{\mu}q_{\mu}\,.
$$

36.2.2   Invariant interval

We have said that the quantity

$$ds^2 = c^2 dt^2 - d\vec{x}^2 \,,$$

interval between two events, is a Lorentz invariant. Let us now consider an event that evolves over time and choose as a reference frame the one where it is at rest. In this kind of reference frame we have

$$d\vec{x}^2 = 0$$

and we obtain

$$ds^2 = c^2 d\tau^2 \,,$$

where we used the greek τ letter for the time measured in the frame where the event is at rest, called also proper time. The quantity on first member is Lorentz invariant, the same must be the quantity on second member. The proper time is related to the time measured in any system (which moves with speed v relative to it) from

$$c^2 dt^2 - d\vec{x}^2 = c^2 d\tau^2 \,,$$

$$c^2 dt^2 - v^2 dt^2 = c^2 d\tau^2 \,,$$

$$(c^2 - v^2) dt^2 = c^2 d\tau^2 \,,$$

$$\left(1 - \frac{v^2}{c^2}\right) dt^2 = d\tau^2 \,,$$

$$\frac{1}{\gamma^2} dt^2 = d\tau^2 \,,$$

$$dt = \gamma \, d\tau \,,$$

where we used the Lorentz factor introduced previously and given by

$$\gamma = \frac{1}{\sqrt{1 - \frac{v^2}{c^2}}} \,.$$

36.2.3 Four-velocity

We define the four-velocity of a particle as

$$u^\mu = \frac{ds^\mu}{d\tau} \,,$$

note that the proper time appears in the definition, with

$$ds^2 = c^2 d\tau^2$$

and

$$dt = \gamma \, d\tau \,.$$

We remember that

$$s^\mu = (ct, x, y, z),$$

so

$$u^\mu = \gamma \frac{ds^\mu}{dt} = (\gamma c, \gamma \vec{v}),$$

where

$$\vec{v} = \left(\frac{dx}{dt}, \frac{dy}{dt}, \frac{dz}{dt} \right)$$

is the velocity of the particle in a certain reference frame S.

Let's calculate the following Lorentz invariant quantity

$$u^\mu u_\mu = \gamma^2 c^2 - \gamma^2 \vec{v}^2 = \gamma^2 (c^2 - \vec{v}^2),$$

by using

$$\gamma = \frac{1}{\sqrt{1 - \frac{v^2}{c^2}}},$$

$$1 - \frac{v^2}{c^2} = \frac{1}{\gamma^2},$$

$$c^2 - v^2 = \frac{c^2}{\gamma^2},$$

we obtain

$$u^\mu u_\mu = \gamma^2 \frac{c^2}{\gamma^2} = c^2.$$

36.2.4 Four-momentum

The relationship between the four-momentum and the four-velocity, for a particle with non zero mass m, is

$$p^\mu = mu^\mu \,.$$

So we can write

$$p^\mu = (m\gamma c, m\gamma \vec{v})$$

and

$$p^\mu p_\mu = m^2 c^2 \,,$$

where

$$\vec{p} = (p_x, p_y, p_z) = m\gamma \vec{v}$$

is the relativistic momentum vector.

The four-momentum of any generic particle (even massless, like the photon) can also be written as

$$p^\mu = \left(\frac{E}{c}, p^1, p^2, p^3\right) = \left(\frac{E}{c}, \vec{p}\right) \,,$$

where E is the energy of the particle.

36.3   Energy and mass-shell relation

36.3.1   Mass-shell relation

From the two formula

$$u^\mu u_\mu = c^2 \,,$$

$$p^\mu = mu^\mu \,,$$

we have calculated

$$p^\mu p_\mu = m^2 u^\mu u_\mu = m^2 c^2 \,.$$

Knowing that the four-momentum can be also written as

$$p^\mu = \left(\frac{E}{c}, p^1, p^2, p^3\right) = \left(\frac{E}{c}, \vec{p}\right) \,,$$

we obtain

$$p^\mu p_\mu = \frac{E^2}{c^2} - \vec{p}^2 \,,$$

from which

$$\frac{E^2}{c^2} - \vec{p}^2 = m^2 c^2 \,,$$

$$E^2 - \vec{p}^2 c^2 = m^2 c^4$$

and the so-called mass-shell relation is obtained

$$E^2 = \vec{p}^2 c^2 + m^2 c^4 \,.$$

From this formula we see, for example, that a massless particle, like the photon, has energy

$$E = |\vec{p}|c \,.$$

36.3.2 Relativistic energy

Comparing the formula

$$p^{\mu} = mu^{\mu} = (m\gamma c, m\gamma \vec{v}) \,,$$

with

$$p^{\mu} = \left(\frac{E}{c}, \vec{p} \right) \,,$$

we obtain the relativistic energy for a particle of non-zero mass m

$$E = mc^2 \gamma$$

and its four-momentum (as previously seen)

$$\vec{p} = m\gamma \vec{v} \,.$$

In the reference frame at rest with the particle (with $v = 0$ and $\gamma = 1$), the energy of a non-zero mass particle becomes

$$E = mc^2 \,,$$

that is the famous formula for the so-called rest energy or mass energy of a particle.

The total energy of a relativistic free particle

$$E = mc^2\gamma \,,$$

can be obtained as the sum of its kinetic energy, called T, and its mass energy

$$E = T + mc^2 \,,$$

hence the kinetic energy is

$$T = E - mc^2 = mc^2\gamma - mc^2 \,,$$

or

$$T = mc^2(\gamma - 1) \,.$$

We now calculate the derivative of the Lorentz factor with respect to the velocity v

$$\gamma = \frac{1}{\sqrt{1 - \frac{v^2}{c^2}}} = \left(1 - \frac{v^2}{c^2}\right)^{-1/2}$$

$$\frac{d\gamma}{dv} = -\frac{1}{2}\left(1 - \frac{v^2}{c^2}\right)^{-3/2}\left(-2\frac{v}{c^2}\right) = \frac{v\gamma^3}{c^2} \,.$$

For the second derivative we have

$$\frac{d^2\gamma}{dv^2} = \frac{d}{dv}\frac{v\gamma^3}{c^2} = \frac{\gamma^3 + 3v\gamma^2 d\gamma/dv}{c^2} = \frac{\gamma^3 + 3v^2\gamma^5/c^2}{c^2}.$$

We consider the following Taylor (McLaurin) series expansion, until the second order, of the Lorentz factor, around $v = 0$

$$\gamma(v) \quad \sim \quad \gamma(0) + v\left(\left.\frac{v\gamma^3}{c^2}\right|_{v=0}\right)$$
$$+ \quad \frac{1}{2}v^2\left(\left.\frac{\gamma^3 + 3v^2\gamma^5/c^2}{c^2}\right|_{v=0}\right),$$

from which

$$\gamma \sim 1 + \frac{1}{2}\frac{v^2}{c^2},$$

valid if

$$\frac{v}{c} \to 0.$$

By inserting this expression in the relativistic kinetic energy

$$T = mc^2(\gamma - 1),$$

we get its well-known non-relativistic expression

$$T_{NR} = mc^2\left(1 + \frac{1}{2}\frac{v^2}{c^2} - 1\right) = \frac{1}{2}mc^2\frac{v^2}{c^2} = \frac{1}{2}mv^2.$$

Chapter 37
Particle decays

37.1 Conservation of four-momentum

In classical physics the laws of conservation of energy and momentum are well known. In relativity these are included into the conservation of the four-momentum which contains both the energy (the temporal component) and the momentum (the three spatial components). We remember also the two expressions for the four-momentum

$$p^\mu = (m\gamma c, m\gamma \vec{v})$$

and

$$p^\mu = \left(\frac{E}{c}, \vec{p}\right) ,$$

with

$$p^2 = p^\mu p_\mu = m^2 c^2$$

and

$$E^2 = \vec{p}^2 c^2 + m^2 c^4 \,.$$

Be careful to not confuse, as already mentioned, the square of p with its contravariant component of index 2, it should be clear from the context which one is the correct interpretation.

37.2 Two-body decay

With the expression two-body decay we denote the decay of a particle into two particles. Let A be the initial particle and B and C those in the final state, we can write schematically

$$A(p) \to B(k) + C(q) \,,$$

where in parentheses are shown the particles four-momenta. The conservation law for the four-momentum reads

$$p^\mu = k^\mu + q^\mu$$

and is valid in any inertial reference frame. In general, it is preferred to write equations that are Lorentz invariant, so that the values are the same in each chosen reference frame.

From the expression below, which is not Lorentz invariant, one can obtain various expressions involving Lorentz invariant quantities. For example, we can consider the square of both members (intended as a scalar product)

$$p^2 = p^\mu p_\mu = (k^\mu + q^\mu)(k_\mu + q_\mu) = (k + q)^2 \,,$$

from which

$$p^\mu p_\mu = k^\mu k_\mu + q^\mu q_\mu + 2k^\mu q_\mu$$

and

$$p^2 = k^2 + q^2 + 2k \cdot q \,,$$

where we have used the identity

$$k^\mu q_\mu = k_\mu q^\mu \,.$$

By remembering that

$$p^2 = p^\mu p_\mu = M^2 c^2 \,,$$
$$k^2 = k^\mu k_\mu = m_1^2 c^2 \,,$$
$$q^2 = q^\mu q_\mu = m_2^2 c^2 \,,$$

where M, m_1, m_2 are, respectively, the mass of the decaying particle, with four-momentum p, and the mass of the two particles produced, with four-momentum k and q.

The conservation of four-momentum therefore implies

$$M^2 c^2 = m_1^2 c^2 + m_2^2 c^2 + 2k \cdot q \,,$$

i.e.

$$k \cdot q = \frac{(M^2 - m_1^2 - m_2^2)c^2}{2} \,,$$

where the scalar product (a Lorentz invariant quantity) at the first member can be calculated in any reference frame, usually one choose the center of mass (CM) system, that is the reference frame where the decaying particle A is at rest.

Other useful Lorentz invariant expressions that can be obtained from

$$p^\mu = k^\mu + q^\mu \,,$$

are the following

$$k^\mu = p^\mu - q^\mu \,,$$
$$k^2 = p^2 + q^2 - 2p \cdot q \,,$$
$$p \cdot q = \frac{(M^2 - m_1^2 + m_2^2)c^2}{2}$$

and

$$q^{\mu} = p^{\mu} - k^{\mu} \, ,$$

$$q^2 = p^2 + k^2 - 2p \cdot k \, ,$$

$$p \cdot k = \frac{(M^2 + m_1^2 - m_2^2)c^2}{2} \, .$$

In summary, the three obtained expressions, that are Lorentz invariant and are related to the scalar products between the four-momenta, are

$$k \cdot q = \frac{(M^2 - m_1^2 - m_2^2)c^2}{2} \, ,$$

$$p \cdot q = \frac{(M^2 - m_1^2 + m_2^2)c^2}{2} \, ,$$

$$p \cdot k = \frac{(M^2 + m_1^2 - m_2^2)c^2}{2} \, .$$

In the case of two-body decay, choosing the CM system to perform the calculations, we can write the three four-momenta in this way

$$p^{\mu} = \left(\frac{E}{c}, \vec{0} \right) \, ,$$

since the particle A is at rest in the CM system and where E is its energy (always in the CM system),

$$k^{\mu} = \left(\frac{E_1}{c}, \vec{k} \right)$$

and

$$q^{\mu} = \left(\frac{E_2}{c}, \vec{q} \right) ,$$

where the energies of the particles and their four-momenta refer to the CM system (which often also coincides with the laboratory system). In the CM system the produced particles (B and C) have opposite momenta, thanks to the momentum conservation law. So we have

$$|\vec{k}| = |\vec{q}| , \quad \vec{k} = -\vec{q},$$

from which

$$k^{\mu} = \left(\frac{E_1}{c}, \vec{k} \right)$$

and

$$q^{\mu} = \left(\frac{E_2}{c}, -\vec{k} \right) .$$

In addition, for the conservation of the temporal component of the four-momentum (energy conservation), we can write

$$\frac{E}{c} = \frac{E_1}{c} + \frac{E_2}{c} ,$$

i.e.

$$E = E_1 + E_2 \,.$$

The energy of the particle A that decays (we are in the CM system) is its rest energy

$$E = Mc^2 \,,$$

so the previous equation becomes

$$E_1 + E_2 = M^2 c^2 \,.$$

In summary, the three four-momenta, in CM system, are

$$p^\mu = \left(Mc, \vec{0} \right) ,$$
$$k^\mu = \left(\frac{E_1}{c}, \vec{k} \right) ,$$
$$q^\mu = \left(\frac{E_2}{c}, -\vec{k} \right)$$

and the three scalar products, calculated before, in the CM system have the form

$$k \cdot q = \frac{E_1 E_2}{c^2} + |\vec{k}|^2 \,,$$
$$p \cdot q = M E_2 \,,$$
$$p \cdot k = M E_1 \,.$$

In particular, from the latter two expressions, using the previous relations, we obtain

$$ME_2 = \frac{(M^2 - m_1^2 + m_2^2)c^2}{2}$$

and

$$ME_1 = \frac{(M^2 + m_1^2 - m_2^2)c^2}{2} \, ,$$

from which the energies in the CM system of the produced particles

$$E_1 = \frac{(M^2 + m_1^2 - m_2^2)c^2}{2M} \, ,$$
$$E_2 = \frac{(M^2 - m_1^2 + m_2^2)c^2}{2M} \, .$$

Analogously, using the other expression

$$\frac{E_1 E_2}{c^2} + |\vec{k}|^2 = \frac{(M^2 - m_1^2 - m_2^2)c^2}{2} \, ,$$
$$|\vec{k}|^2 = \frac{(M^2 - m_1^2 - m_2^2)c^2}{2} - \frac{E_1 E_2}{c^2} \, .$$

Moreover

$$|\vec{k}|^2 = \frac{(M^2 - m_1^2 - m_2^2)c^2}{2} - \frac{(M^2 + m_1^2 - m_2^2)(M^2 - m_1^2 + m_2^2)c^2}{4M^2} \, ,$$

from which

$$|\vec{k}|^2 = \frac{[M^4 + (m_1^2 - m_2^2)^2 - 2M^2(m_1^2 + m_2^2)]c^2}{4M^2},$$

$$|\vec{k}| = \frac{\sqrt{M^4 + (m_1^2 - m_2^2)^2 - 2M^2(m_1^2 + m_2^2)}\,c}{2M}$$

and this is the expression, in the CM system, of the modulus of momentum of the produced particles.

The total energy of a free particle is given, as already said, by the sum of its rest energy and its kinetic energy. Therefore we have

$$T_1 + m_1c^2 = E_1 = \frac{(M^2 + m_1^2 - m_2^2)c^2}{2M},$$

$$T_2 + m_2c^2 = E_2 = \frac{(M^2 - m_1^2 + m_2^2)c^2}{2M},$$

from which

$$T_1 = \frac{(M^2 + m_1^2 - m_2^2)c^2}{2M} - m_1c^2,$$

$$T_2 = \frac{(M^2 - m_1^2 + m_2^2)c^2}{2M} - m_2c^2$$

and

$$T_1 = \frac{(M^2 + m_1^2 - m_2^2 - 2Mm_1)c^2}{2M},$$

$$T_2 = \frac{(M^2 - m_1^2 + m_2^2 - 2Mm_2)c^2}{2M}.$$

Finally

$$T_1 = \frac{[(M - m_1)^2 - m_2^2]c^2}{2M},$$

$$T_2 = \frac{[(M - m_2)^2 - m_1^2]c^2}{2M}.$$

37.3 Impossibility of the decay of a free photon

Consider the decay of a photon into two particles B and C

$$\gamma(p) \to B(k) + C(q),$$

where in parenthesis are shown the four-momenta of the particles, with the same notation used previously when we discussed the two-body decay. The photon is a massless particle, $M = 0$, so its four-momentum has the form

$$p^\mu = \left(\frac{E}{c}, \vec{p}\right) = (|\vec{p}|, \vec{p}),$$

where we have used the mass-shell relation

$$E^2 = \vec{p}^2 c^2 + M^2 c^4 = \vec{p}^2 c^2 = |\vec{p}| c.$$

By choosing the CM system where

$$\vec{k} = -\vec{q},$$

we have

$$k^\mu = \left(\frac{E_1}{c}, \vec{k}\right),$$
$$q^\mu = \left(\frac{E_2}{c}, -\vec{k}\right)$$

and therefore from the four-momentum conservation law

$$p^\mu = k^\mu + q^\mu$$

you would have

$$\begin{cases} |\vec{p}| = \frac{E_1}{c} + \frac{E_2}{c} > 0 \\ \vec{p} = \vec{0} \end{cases},$$

which is absurd.

37.4 Muon decay

Consider the decay of the muon

$$\mu^-(P) \rightarrow e^-(p) + \overline{\nu}_e(q) + \nu_\mu(k),$$

that is a three-body decay, where two of the three final particles (the muon neutrino, with four-momentum k, and the electron antineutrino, with four-momentum q) can be considered massless. We write the four-momenta in the

277

CM system, where the muon is at rest

$$P^\mu = \left(m_\mu c, \vec{0}\right),$$

$$p^\mu = \left(\frac{E_e}{c}, \vec{p}\right),$$

$$q^\mu = \left(\frac{E_{\bar{\nu}}}{c}, \vec{q}\right),$$

$$k^\mu = \left(\frac{E_\nu}{c}, \vec{k}\right).$$

We are interested in the maximum value of the kinetic energy of the electron in the CM system.

Meanwhile, for the momentum conservation law

$$\vec{p} + \vec{q} + \vec{k} = \vec{0},$$

from which

$$\vec{q} + \vec{k} = -\vec{p}.$$

The four-momentum conservation law is

$$P^\mu = p^\mu + k^\mu + q^\mu,$$

i.e.

$$P^\mu - p^\mu = k^\mu + q^\mu.$$

Squaring both members (scalar products)

$$(P^\mu - p^\mu)(P_\mu - p_\mu) = (k^\mu + q^\mu)(k_\mu + q_\mu) \,,$$
$$P^2 + p^2 - 2P \cdot p = (k + q)^2 \,,$$

from which, knowing that

$$P^2 = m_\mu^2 c^2 \,, \quad p^2 = m_e^2 c^2$$

we obtain

$$m_\mu^2 c^2 + m_e^2 c^2 - 2P \cdot p = (k + q)^2 \,.$$

By remembering

$$P^\mu = (m_\mu c, \vec{0}) \,,$$
$$p^\mu = \left(\frac{E_e}{c}, \vec{p} \right) \,,$$

we obtain

$$P \cdot p = m_\mu E_e \,,$$

from which

$$m_\mu^2 c^2 + m_e^2 c^2 - 2m_\mu E_e = (k + q)^2$$

and

$$2m_\mu E_e = m_\mu^2 c^2 + m_e^2 c^2 - (k + q)^2 \,,$$

or

$$E_e = \frac{(m_\mu^2 + m_e^2)c^2 - (k+q)^2}{2m_\mu}.$$

Let's consider the four-momentum

$$k^\mu + p^\mu = \left(\frac{E_\nu}{c} + \frac{E_{\bar\nu}}{c}, \vec{k} + \vec{q}\right),$$

using

$$\vec{q} + \vec{k} = -\vec{p},$$

we have

$$k^\mu + p^\mu = \left(E_\nu/c + E_{\bar\nu}/c, -\vec{p}\right),$$

so that

$$(k+q)^2 = \left(\frac{E_\nu}{c} + \frac{E_{\bar\nu}}{c}\right)^2 + \vec{p}^2 > 0.$$

From the equation

$$E_e = \frac{(m_\mu^2 + m_e^2)c^2 - (k+q)^2}{2m_\mu},$$

we can say that this quantity (electron energy in the CM system) reaches the maximum if the quantity

$$(k+q)^2$$

is minimal. We calculate

$$(k + q)^2 = k^2 + q^2 + 2k \cdot q = 2k \cdot q = \frac{E_\nu E_{\overline{\nu}}}{c^2} - \vec{k} \cdot \vec{q}$$

and, using the mass-shell relation for neutrinos, considered as massless particles,

$$E_\nu = |\vec{k}|c, \quad E_{\overline{\nu}} = |\vec{q}|c,$$

we obtain

$$(k + q)^2 = \frac{E_\nu E_{\overline{\nu}}}{c^2} - \frac{E_\nu E_{\overline{\nu}}}{c^2} \cos \theta$$

e

$$(k + q)^2 = \frac{E_\nu E_{\overline{\nu}}}{c^2} (1 - \cos \theta)$$

where θ is the angle between the neutrinos four-momenta in the CM system. It is easy to verify that the lower limit for this quantity is zero.

So, using the equation

$$E_e = \frac{(m_\mu^2 + m_e^2)c^2 - (k + q)^2}{2m_\mu},$$

we can write

$$E_e < \frac{(m_\mu^2 + m_e^2)c^2}{2m_\mu},$$

from which, for the kinetic energy of the electron, we obtain

$$T_e = E_e - m_e c^2$$

and

$$T_e < \frac{(m_\mu^2 + m_e^2)c^2}{2m_\mu} - m_e c^2 \,,$$
$$T_e < \frac{(m_\mu^2 + m_e^2 - 2m_e m_\mu)c^2}{2m_\mu}$$

and, finally,

$$T_e < \frac{(m_\mu - m_e)^2 c^2}{2m_\mu} \,,$$

that can be considered as the maximum value for the kinetic energy of the electron in the muon decay.

Chapter 38

Electric and magnetic fields

38.1 Electromagnetic tensor

In relativity the electric and magnetic fields

$$\vec{E} = (E_x, E_y, E_z) \,,$$
$$\vec{B} = (B_x, B_y, B_z) \,,$$

are not separately Lorentz invariant quantities. They appear in the so-called electromagnetic tensor, represented by the following matrix

$$F^{\mu\nu} = \begin{pmatrix} 0 & -E_x/c & -E_y/c & -E_z/c \\ E_x/c & 0 & -B_z & B_y \\ E_y/c & B_z & 0 & -B_x \\ E_z/c & -B_y & B_x & 0 \end{pmatrix} \,,$$

where it is possible to observe, as elements, the components of the electric and magnetic fields.

38.1.1 A Lorentz invariant quantity

The following quantity

$$F^{\mu\nu}F_{\mu\nu} = F^{\mu\nu}\eta_{\mu\alpha}\eta_{\nu\beta}F^{\alpha\beta}$$

is a Lorentz scalar since all the indices are summed over. Moreover, we observe that the electromagnetic tensor is antisymmetric in the exchange of the two indices (exchange of rows and columns in the matrix), i.e.

$$F^{\mu\nu} = -F^{\nu\mu} \,,$$

hence the previous equation, using also the symmetry of the metric tensor

$$\eta_{\mu\alpha} = \eta_{\alpha\mu}$$

becomes

$$F^{\mu\nu}F_{\mu\nu} = F^{\mu\nu}\eta_{\mu\alpha}\eta_{\nu\beta}F^{\alpha\beta} = -F^{\mu\nu}\eta_{\nu\beta}F^{\beta\alpha}\eta_{\alpha\mu} \,.$$

Since all the indices are summed over, the last member of the equation can be seen as the trace of a product of matrices, in this way

$$F^{\mu\nu}F_{\mu\nu} = -\operatorname{Tr}\left(F\eta F\eta\right) \,.$$

Note that in the previous formula the factors can be moved using the commutative property, being the single elements of matrices, while in the latter formula the product between matrices in the trace is not commutative and one can only use the property of the trace of a matrices product. This allows us to move the matrices cyclically, for example given four matrices A, B, C and D we have the following identity

$$\begin{aligned} \text{Tr}\,(ABCD) &= \text{Tr}\,(BCDA) = \text{Tr}\,(CDAB) \\ &= \text{Tr}\,(DABC)\,. \end{aligned}$$

We recall that the product between two matrices A and B is a matrix with elements given by

$$(AB)_{ik} = \sum_j A_{ij} B_{jk}\,,$$

while the trace of a matrix A is given by

$$\text{Tr}\,(A) = \sum_i A_{ii}\,.$$

From the previous equation we obtain

$$F^{\mu\nu} F_{\mu\nu} = -\text{Tr}\,(F\eta F\eta) = -2(\vec{E}^2 - \vec{B}^2)\,,$$

which is a Lorentz invariant quantity and has the same value in all inertial reference frames.

38.1.2 Transformation of the fields $\vec{E}$ and $\vec{B}$

The electromagnetic tensor

$$F^{\mu\nu} = \begin{pmatrix} 0 & -E_x/c & -E_y/c & -E_z/c \\ E_x/c & 0 & -B_z & B_y \\ E_y/c & B_z & 0 & -B_x \\ E_z/c & -B_y & B_x & 0 \end{pmatrix},$$

calculated in another reference frame is obtained by applying the transformation matrices, i.e.

$$F'^{\mu\nu} = \Lambda^\mu{}_\alpha \Lambda^\nu{}_\beta F^{\alpha\beta}.$$

Suppose we have a charge q in motion with speed v along the abscissa axis in an inertial reference frame S. We can introduce the system S' where the charge q is at rest. This reference frame is in motion with speed v with respect to S along the abscissa axis (with parallel axes and coinciding origins at the initial time). We know that in this case the transformation matrix between the two systems has the form

$$\Lambda^\mu{}_\nu = \begin{pmatrix} \gamma & -v\gamma/c & 0 & 0 \\ -v\gamma/c & \gamma & 0 & 0 \\ 0 & 0 & 1 & 0 \\ 0 & 0 & 0 & 1 \end{pmatrix},$$

which can be written also as

$$\Lambda^{\mu}{}_{\nu} = \begin{pmatrix} \gamma & -\gamma\beta & 0 & 0 \\ -\gamma\beta & \gamma & 0 & 0 \\ 0 & 0 & 1 & 0 \\ 0 & 0 & 0 & 1 \end{pmatrix},$$

having introduced the quantity

$$\vec{\beta} = \frac{\vec{v}}{c}, \quad |\vec{\beta}| = \beta = \frac{v}{c}.$$

In the reference frame S' the charge q is at rest and it generates only an electric field (and no magnetic field) given by

$$\vec{E}' = \frac{kq}{r'^2}\hat{r}' = \frac{kq}{(x'^2 + y'^2 + z'^2)^{3/2}} \left(x', y', z'\right),$$

where we have used

$$\hat{r}' = \frac{\vec{r}'}{r} = \frac{(x', y', z')}{r}.$$

The electromagnetic tensor in this case has the form

$$F'^{\mu\nu} = \begin{pmatrix} 0 & -E'_x/c & -E'_y/c & -E'_z/c \\ E'_x/c & 0 & 0 & 0 \\ E'_y/c & 0 & 0 & 0 \\ E'_z/c & 0 & 0 & 0 \end{pmatrix}.$$

From the equation

$$F'^{\mu\nu} = \Lambda^{\mu}_{\ \alpha}\Lambda^{\nu}_{\ \beta}F^{\alpha\beta},$$

we have

$$F'^{\mu\nu} = \Lambda^{\mu}_{\ \alpha}F^{\alpha\beta}\Lambda^{\nu}_{\ \beta} = \Lambda^{\mu}_{\ \alpha}F^{\alpha\beta}(\Lambda^{T})_{\beta}^{\ \nu}$$

and, in matrix form,

$$F' = \Lambda F\Lambda^{T},$$

from which, applying appropriately inverse matrices,

$$\Lambda^{-1}\Lambda F\Lambda^{T}(\Lambda^{T})^{-1} = \Lambda^{-1}F'(\Lambda^{T})^{-1},$$
$$F = \Lambda^{-1}F'(\Lambda^{T})^{-1} = \Lambda^{-1}F'(\Lambda^{-1})^{T},$$

in fact

$$(\Lambda^{-1})^{T} = (\Lambda^{T})^{-1}.$$

We calculate the inverse of the Λ matrix

$$\Lambda^{-1} = \begin{pmatrix} \gamma & \gamma\beta & 0 & 0 \\ \gamma\beta & \gamma & 0 & 0 \\ 0 & 0 & 1 & 0 \\ 0 & 0 & 0 & 1 \end{pmatrix}$$

and the inverse of the transpose which coincides with the transpose of the inverse

$$\left(\Lambda^T\right)^{-1} = \begin{pmatrix} \gamma & \gamma\beta & 0 & 0 \\ \gamma\beta & \gamma & 0 & 0 \\ 0 & 0 & 1 & 0 \\ 0 & 0 & 0 & 1 \end{pmatrix},$$

from which

$$\Lambda^{-1}F' = \begin{pmatrix} \gamma & \gamma\beta & 0 & 0 \\ \gamma\beta & \gamma & 0 & 0 \\ 0 & 0 & 1 & 0 \\ 0 & 0 & 0 & 1 \end{pmatrix} \cdot \begin{pmatrix} 0 & -\frac{E'_x}{c} & -\frac{E'_y}{c} & -\frac{E'_z}{c} \\ \frac{E'_x}{c} & 0 & 0 & 0 \\ \frac{E'_y}{c} & 0 & 0 & 0 \\ \frac{E'_z}{c} & 0 & 0 & 0 \end{pmatrix}$$

$$= \begin{pmatrix} \gamma\beta\frac{E'_x}{c} & -\gamma\frac{E'_x}{c} & -\gamma\frac{E'_y}{c} & -\gamma\frac{E'_z}{c} \\ \gamma\frac{E'_x}{c} & -\gamma\beta\frac{E'_x}{c} & -\gamma\beta\frac{E'_y}{c} & -\gamma\beta\frac{E'_z}{c} \\ \frac{E'_y}{c} & 0 & 0 & 0 \\ \frac{E'_z}{c} & 0 & 0 & 0 \end{pmatrix}.$$

From which, doing the calculations,

$$\Lambda^{-1}F'\left(\Lambda^T\right)^{-1} = \begin{pmatrix} 0 & -\frac{E'_x}{c} & -\gamma\frac{E'_y}{c} & -\gamma\frac{E'_z}{c} \\ \frac{E'_x}{c} & 0 & -\gamma\beta\frac{E'_y}{c} & -\gamma\beta\frac{E'_z}{c} \\ \gamma\frac{E'_y}{c} & \gamma\beta\frac{E'_y}{c} & 0 & 0 \\ \gamma\frac{E'_z}{c} & \gamma\beta\frac{E'_z}{c} & 0 & 0 \end{pmatrix}.$$

Starting from

$$F = \Lambda^{-1}F'\left(\Lambda^{-1}\right)^T,$$

with

$$
F = \begin{pmatrix} 0 & -E_x/c & -E_y/c & -E_z/c \\ E_x/c & 0 & -B_z & B_y \\ E_y/c & B_z & 0 & -B_x \\ E_z/c & -B_y & B_x & 0 \end{pmatrix},
$$

we have

$$
\begin{pmatrix} 0 & -E_x/c & -E_y/c & -E_z/c \\ E_x/c & 0 & -B_z & B_y \\ E_y/c & B_z & 0 & -B_x \\ E_z/c & -B_y & B_x & 0 \end{pmatrix}
$$
$$
= \begin{pmatrix} 0 & -E'_x/c & -\gamma E'_y/c & -\gamma E'_z/c \\ E'_x/c & 0 & -\gamma\beta E'_y/c & -\gamma\beta E'_z/c \\ \gamma E'_y/c & \gamma\beta E'_y/c & 0 & 0 \\ \gamma E'_z/c & \gamma\beta E'_z/c & 0 & 0 \end{pmatrix}.
$$

By comparing component by component, we obtain the following relationships between the electric and magnetic fields in the two reference frame S and S'

$$
E_x = E'_x\,, \quad E_y = \gamma E'_y\,, \quad E_z = \gamma E'_z\,,
$$
$$
B_x = 0\,, \quad B_y = -\gamma\beta\frac{E'_z}{c} \quad B_z = \gamma\beta\frac{E'_y}{c}\,.
$$

To obtain the corresponding expressions in the S frame, where the charge is in motion with velocity v, further calculations are needed. Indeed, we recall the expression

$$\vec{E}' = \frac{kq}{(x'^2 + y'^2 + z'^2)^{3/2}} \left(x', y', z' \right),$$

from which

$$E_x = E_x' = \frac{kqx'}{(x'^2 + y'^2 + z'^2)^{3/2}},$$

$$E_y = \gamma E_y' = \frac{kq\gamma y'}{(x'^2 + y'^2 + z'^2)^{3/2}},$$

$$E_z = \gamma E_z' = \frac{kq\gamma z'}{(x'^2 + y'^2 + z'^2)^{3/2}}$$

e

$$B_x = 0,$$

$$B_y = -\gamma\beta\frac{E_z'}{c} = -\frac{kq\gamma\beta z'}{c(x'^2 + y'^2 + z'^2)^{3/2}},$$

$$B_z = \gamma\beta\frac{E_y'}{c} = \frac{kq\gamma\beta y'}{c(x'^2 + y'^2 + z'^2)^{3/2}}.$$

In particular we have to write all the quantities with the corresponding value in the frame S, so we use the Lorentz transformations

$$s'^\mu = \Lambda^\mu_{\ \nu} s^\nu$$

where, in this case,

$$s^\mu = (ct, x, y, z)$$

e

$$\begin{cases} x' = \gamma(x - vt) \\ y' = y \\ z' = z \\ t' = \gamma\left(t - \frac{xv}{c^2}\right) \end{cases}$$

The previous equations become

$$E_x = \frac{kq\gamma(x - vt)}{\left([\gamma(x - vt)]^2 + y^2 + z^2\right)^{3/2}},$$

$$E_y = \frac{kq\gamma y}{\left([\gamma(x - vt)]^2 + y^2 + z^2\right)^{3/2}},$$

$$E_z = \frac{kq\gamma z}{\left([\gamma(x - vt)]^2 + y^2 + z^2\right)^{3/2}}$$

e

$$B_x = 0\,,$$

$$B_y = -\frac{kq\gamma\beta z}{c\left([\gamma(x - vt)]^2 + y^2 + z^2\right)^{3/2}},$$

$$B_z = \frac{kq\gamma\beta y}{c\left([\gamma(x - vt)]^2 + y'^2 + z'^2\right)^{3/2}}\,.$$

The electric field E in the frame S is

$$\vec{E} = \frac{kq\gamma\left(x - vt, y, z\right)}{\left([\gamma(x - vt)]^2 + y^2 + z^2\right)^{3/2}},$$

while the magnetic field is

$$\vec{B} = \frac{kq\gamma\beta\left(0, y, z\right)}{c\left([\gamma(x - vt)]^2 + y^2 + z^2\right)^{3/2}} \cdot$$

Moreover, we can write the Lorentz factor

$$\gamma = \frac{1}{\sqrt{1 - \frac{v^2}{c^2}}}$$

as a function of β

$$\beta = \frac{v}{c}$$

as

$$\gamma = \frac{1}{\sqrt{1 - \beta^2}},$$

from which

$$\gamma = \frac{1}{\sqrt{1 - \beta^2}},$$

$$\beta = \frac{\sqrt{\gamma^2 - 1}}{\gamma},$$

$$\gamma\beta = \sqrt{\gamma^2 - 1}.$$

Finally

$$\vec{E} = \frac{kq\gamma\big(x - vt, y, z\big)}{\big([\gamma(x - vt)]^2 + y^2 + z^2\big)^{3/2}} \, ,$$

$$\vec{B} = \frac{kq\sqrt{\gamma^2 - 1}\,(0, y, z)}{c\big([\gamma(x - vt)]^2 + y^2 + z^2\big)^{3/2}} \, .$$

We observe that in an inertial reference frame S where a charge q is in motion with velocity v it generates also a magnetic field, not present in the system S' where the charge is at rest. Furthermore, the electric and magnetic fields in the frame S are also functions of time.

Chapter 39

Equations of motion

The equations of motion regard both the motion of charges subjected to electric and magnetic fields and the motion of fields in the presence of charges and sources. A rigorous discussion starts from the Lagrangian of a relativistic particle in the presence of an electromagnetic field to obtain the equations of motion, which leads to the Lorentz equation. For the others it is necessary to start from the Lagrangian of the electromagnetic field and then obtain the Maxwell's equations. In this chapter we will limit ourselves to expose the equations rather than deriving them formally from the beginning, since the concepts of Lagrangian, Euler-Lagrange equations and four-potential should be introduced.

39.1 Lorentz equation

A particle with charge q and velocity v, in presence of an electric field $\vec{E}$ and a magnetic field $\vec{B}$, suffers the Lorentz force

$$\vec{F}_L = q\vec{E} + q\vec{v} \times \vec{B}\,,$$

with equation of motion

$$\frac{d(m\gamma\vec{v})}{dt} = q\vec{E} + q\vec{v} \times \vec{B}\,.$$

The equations of motion in covariant form are written as

$$\frac{dp^\mu}{d\tau} = qu_\nu F^{\nu\mu}\,,$$

in fact

$$dt = \gamma\,d\tau$$

and

$$qu_\nu F^{\nu 0} = \frac{q\gamma}{c}\,\vec{v} \cdot \vec{E}\,,$$

$$qu_\nu F^{\nu j} = qE_j + q\gamma(\vec{v} \times \vec{B})_j\,.$$

The first of these two equations provides the additional equation of motion

$$\frac{dp^0}{d\tau} = \frac{q\gamma}{c}\,\vec{v} \cdot \vec{E}\,.$$

39.2 Maxwell's equations

The Maxwell's equations in vacuum, in local form, are

$$\begin{cases} \vec{\nabla} \cdot \vec{E} = \frac{\rho}{\epsilon_0} \\ \vec{\nabla} \cdot \vec{B} = 0 \\ \vec{\nabla} \times \vec{E} = -\frac{\partial \vec{B}}{\partial t} \\ \vec{\nabla} \times \vec{B} = \mu_0 \vec{J} + \epsilon_0 \mu_0 \frac{\partial \vec{E}}{\partial t} \end{cases},$$

with

$$c = \frac{1}{\sqrt{\epsilon_0 \mu_0}},$$

where $\vec{E}$ and $\vec{B}$ are the electric and magnetic fields, ϵ_0 is the vacuum permeability constant, μ_0 is the vacuum magnetic permeability constant, ρ is the charge density and J is the current density vector.

In the absence of sources, i.e. with zero charge density and zero current density, Maxwell's equations take the form

$$\begin{cases} \vec{\nabla} \cdot \vec{E} = 0 \\ \vec{\nabla} \cdot \vec{B} = 0 \\ \vec{\nabla} \times \vec{E} = -\frac{\partial \vec{B}}{\partial t} \\ \vec{\nabla} \times \vec{B} = \epsilon_0 \mu_0 \frac{\partial \vec{E}}{\partial t} \end{cases}.$$

Two of the four Maxwell's equations in vacuum, in covariant form, are written as

$$\partial_\mu F^{\mu\nu} = \mu_0 J^\nu \,,$$

where in the second member is shown the four-current

$$J^\mu = (c\rho, \vec{J}) \,,$$

where we have used the electromagnetic tensor

$$F^{\mu\nu} = \begin{pmatrix} 0 & -E_x/c & -E_y/c & -E_z/c \\ E_x/c & 0 & -B_z & B_y \\ E_y/c & B_z & 0 & -B_x \\ E_z/c & -B_y & B_x & 0 \end{pmatrix}$$

and where we have introduced the four-divergence defined as

$$\partial_\mu = \frac{\partial}{\partial s^\mu} = \left(\frac{1}{c}\frac{\partial}{\partial t}, \frac{\partial}{\partial x}, \frac{\partial}{\partial y}, \frac{\partial}{\partial z} \right) \,.$$

The other two Maxwell's equations are written as

$$\partial_\mu F_{\nu\sigma} + \partial_\nu F_{\sigma\mu} + \partial_\sigma F_{\mu\nu} = 0 \,,$$

note that in this formula no index is summed over.

As an example, we also report some steps to pass from the covariant formulation to the traditional one with the electric and magnetic field vectors

$$\partial_\mu F^{\mu 0} = \frac{1}{c}\, \vec{\nabla} \cdot \vec{E}\,,$$
$$\partial_\mu F^{\mu j} = (\vec{\nabla} \times \vec{B})_j - \frac{1}{c^2}\frac{\partial E_j}{\partial t}\,.$$

Chapter 40

Electromagnetic waves

40.1 Wave equation

We consider the Maxwell's third equation in vacuum and in the absence of sources

$$\vec{\nabla} \times \vec{E} = -\frac{\partial \vec{B}}{\partial t}$$

and we calculate the curl of both members

$$\vec{\nabla} \times \vec{\nabla} \times \vec{E} = -\frac{\partial \vec{\nabla} \times \vec{B}}{\partial t} \, .$$

Using Maxwell's fourth equation

$$\vec{\nabla} \times \vec{B} = \epsilon_0 \mu_0 \frac{\partial \vec{E}}{\partial t} \, ,$$

we have

$$\vec{\nabla} \times \vec{\nabla} \times \vec{E} = -\frac{\partial}{\partial t}\left(\epsilon_0 \mu_0 \frac{\partial \vec{E}}{\partial t}\right),$$

$$\vec{\nabla} \times \vec{\nabla} \times \vec{E} = -\epsilon_0 \mu_0 \frac{\partial^2 \vec{E}}{\partial t^2}.$$

The curl of the curl of $\vec{E}$ is given by

$$\vec{\nabla} \times \vec{\nabla} \times \vec{E} = \vec{\nabla}(\vec{\nabla} \cdot \vec{E}) - \vec{\nabla}^2 \vec{E}.$$

From the first Maxwell equation

$$\vec{\nabla} \cdot \vec{E} = 0,$$

from which

$$\vec{\nabla} \times \vec{\nabla} \times \vec{E} = -\vec{\nabla}^2 \vec{E}$$

and the equation becomes

$$\vec{\nabla}^2 \vec{E} = \epsilon_0 \mu_0 \frac{\partial^2 \vec{E}}{\partial t^2},$$

i.e.

$$\vec{\nabla}^2 \vec{E} = \frac{1}{c^2}\frac{\partial^2 \vec{E}}{\partial t^2}.$$

This is properly the form of the wave equation for the electric field, which propagates at the speed of light c. Similarly, we consider the Maxwell's fourth equation, in vacuum and in the absence of sources,

$$\vec{\nabla} \times \vec{B} = \epsilon_0 \mu_0 \frac{\partial \vec{E}}{\partial t}$$

and we calculate the curl of both members

$$\vec{\nabla} \times \vec{\nabla} \times \vec{B} = \epsilon_0 \mu_0 \frac{\partial \vec{\nabla} \times \vec{E}}{\partial t}\,.$$

Using Maxwell's third equation

$$\vec{\nabla} \times \vec{E} = -\frac{\partial \vec{B}}{\partial t}$$

we have

$$\vec{\nabla} \times \vec{\nabla} \times \vec{B} = -\epsilon_0 \mu_0 \frac{\partial^2 \vec{B}}{\partial t^2}\,.$$

The curl of the curl of $\vec{B}$ is given by

$$\vec{\nabla} \times \vec{\nabla} \times \vec{B} = \vec{\nabla}(\vec{\nabla} \cdot \vec{B}) - \vec{\nabla}^2 \vec{B}\,.$$

From Maxwell's second equation

$$\vec{\nabla} \cdot \vec{B} = 0\,,$$

from which

$$\vec{\nabla} \times \vec{\nabla} \times \vec{B} = -\vec{\nabla}^2 \vec{B}$$

and the equation becomes

$$\vec{\nabla}^2 \vec{B} = \epsilon_0 \mu_0 \frac{\partial^2 \vec{B}}{\partial t^2} \, ,$$

i.e.

$$\vec{\nabla}^2 \vec{B} = \frac{1}{c^2} \frac{\partial^2 \vec{B}}{\partial t^2} \, .$$

This is, as for the case of the electric field, the form of the wave equation for the magnetic field, which propagates at the speed of light c.

We have seen that in the vacuum and in the absence of sources, the electric field and the magnetic field propagate like waves at the speed of light, the so-called electromagnetic waves.

Books included in this series

Complex numbers
ISBN: 9798674312185

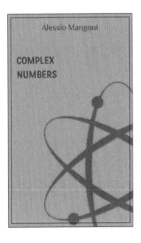

Special relativity

ISBN: 9798675703647

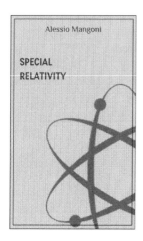

The mathematics of quantum mechanics

ISBN: 9798645275037

The Dirac equation
ISBN: 9798666724644

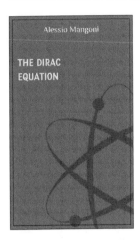

Relativity, decays and electromagnetic fields
ISBN: 9798663840200

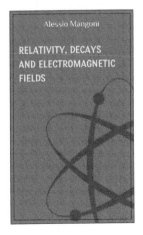

Other books of the author

Fundamentals of physics

ISBN: 9798655711945

Manufactured by Amazon.ca
Bolton, ON

26566914R00171